Water Conservation

for Small- and Medium-Sized Utilities

Water Conservation

for Small- and Medium-Sized Utilities

Deborah Green

Water Conservation for Small- and Medium-Sized Utilities

Disclaimer
This handbook is provided for informational purposes only, with the understanding that the publisher, editors, and authors are not thereby engaged in rendering engineering or other professional services. The authors, editors, and publisher make no claim as to the accuracy of the handbook's contents, or their applicability to any particular circumstance. The editors, authors, and publisher accept no liability to any person for the information or advice provided in this book or for loss or damages incurred by any person as a result of reliance on its contents. The reader is urged to consult with an appropriate licensed professional before taking any action or making any interpretation that is within the realm of a licensed professional practice.

AWWA Publications Manager: Gay Porter De Nileon
Project Manager/Technical Editor: Melissa Valentine
Production Editor: Cheryl Armstrong

Library of Congress Cataloging-in-Publication Data
Green, Deborah, 1950-
Water conservation guidebook for small- and medium-sized utilities / by Deborah Green.
p. cm.
Includes bibliographical references and index.
ISBN 978-1-58321-746-7 (alk. paper)
1. Water conservation. I. Title.
TD388.G729 2010
363.6'10682--dc22
2010001652

ISBN 1-58321-746-0
9781583217467

6666 West Quincy Avenue
Denver, CO 80235-3098
303.794.7711
www.awwa.org

Printed on Recycled Paper

Contents

Review Consumer Conservation Measures 33

Potential Water Savings 83

Designing a Conservation Program 93

Implementing a Conservation Program 105

Appendix A 109

Appendix B 117

Acknowledgments

In 2007, the American Water Works Association (AWWA) Water Conservation Division Planning Evaluation and Research Committee proposed updating a 1993 guide published by the Pacific Northwest Section of the American Water Works Association (PNWS). The result is this book, which was approved by the AWWA Technical Resources Committee and AWWA Water Conservation Division, and supported by AWWA Publications. The lead revision writer was Deborah Green, Water Media Services, with guidance by Judi Ranton, Portland Water Bureau. Reviewers were William Maddaus, Alice Darilek, Elizabeth Gardener, Paul Lander, and Mary Ann Dickinson. Those contributing specific information included Michelle Maddaus, Peter Mayer, John Koeller, and William Gauley.

The orginal version of this book was produced with a grant from the US Environmental Protection Agency and selected by USEPA's Water Use Efficiency Task Force as one of 12 water conservation projects funded during fiscal year 1991. Bryan Yim, USEPA Region 10, served as project officer for USEPA Grant X000576-01-0. The original version was prepared by William Maddaus of Montgomery Watson, under the direction of the members of the PNWS Water Conservation Committee, whose members included Marla Carter, Jack Donohue, Jane Evancho, Kelly Lange, Jerry Parker, Mark Spahr, Lisa Tobin, Linda Utgard, and Tiffany Yelton. The project manager was Cynthia Dietz. In addition to Maddaus, Montgomery Watson's project team included Peter Macy, Lisa Obermeyer, Nancy Patania, and Edward Tenney. Other contributors to the 1993 publication were Annabel DeLong, Noel Groshong, and Chrys Uhlig.

Foreword

HOW TO USE THIS GUIDEBOOK

This guidebook is written for the small and medium-sized water utility that might be facing water shortage issues and is therefore considering implementation of a water conservation program. As defined by the US Environmental Protection Agency (USEPA), a small utility has fewer than 10,000 service connections; a medium utility has between 10,000 and 100,000 connections.

With a menu of possible conservation techniques and approaches from which to choose, this guidebook can be used by small and medium-sized utilities in all parts of the country, operating under their own unique circumstances. Special worksheets have been prepared to guide the utility through the planning and implementation process, with special attention to the needs and issues of small to medium-sized systems.

Small utilities should use the material on conservation measures (Chapter 4), plus fill out the worksheets that seem appropriate and helpful and for which data is readily available. That would include worksheets 1 and 2A in Chapter 2.

Medium-sized utilities should attempt to fill out worksheets 1, 2A, 2B, 2C and 2D in Chapter 2. Although this guidebook does not target large utilities, those large utilities just starting a program can also productively fill out the same worksheets as a medium-sized utility plus use the material in AWWA M52 *Water Conservation Programs—A Planning Manual* to perform a detailed benefit-cost analysis on alternative conservation measures.

ORGANIZATION OF THE GUIDEBOOK

Chapter 1: Introduction, defines water conservation and the kinds of conservation activity addressed in the guidebook. It discusses the reasons for conserving water and the possible effects conservation may have from a variety of perspectives.

Chapter 2: Evaluate the Utility's System, shows how to put together information on utility supply and demand characteristics, as related to choice of water conservation approaches. Worksheets are provided. This information will be used in Chapter 6, which discusses how to design a conservation program.

Chapter 3: Review Utility Conservation Measures and **Chapter 4: Review Consumer Conservation Measures,** give an overview of potential measures that can be part of a conservation program. Each type of measure is described in terms of general feasibility, costs, benefits, and other considerations.

Chapter 5: Consider Potential Water Savings, summarizes the overall water savings that can be achieved by applying the types of measure described in Chapters 3 and 4. An example program for small utilities, which combines a number of conservation measures, is described.

Chapter 6: Design a Conservation Program, includes ways to define the utility's program goals, select appropriate conservation measures, and estimate benefits and costs. Worksheets are included to help evaluate potential conservation measures.

Chapter 7: Implementing the Conservation Program, discusses the process required to put the program into place, including evaluation.

The **Appendices** contain material that supplements the information in Chapters 1 through 7.

The **Glossary** is provided to help use this guidebook.

The **References** identify sources that are referred to in the text and that can provide more detailed information about conservation measures and methods.

An attempt has been made to keep the planning process straightforward and within the resources of a smaller utility. It must be said at the outset that once the commitment has been made to increase a utility's water conservation effort, a source of funding needs to be identified as well as a *point person* designated, referred to as the Water Conservation Coordinator.

A minimal program could start with as little as $1,000 for funding. In the minimum program, the water conservation coordinator will have other duties. The minimum funding budget of $1,000–$10,000 is for projects and does not include salary or overhead. A moderate program could have a full-time water conservation coordinator and a project budget of $10,000–$100,000 but may still not require more than a half-time water conservation coordinator. A maximum program could have several staff members working on water conservation and a program budget of more than $100,000. Chapter 3 includes suggestions on sources of funding.

Improved Resources Available for Utility Water Conservation Programs

The Pacific Northwest Section American Water Works Association published the first edition of this Guidebook in 1993. Since this time, many improved resources have been made available for utilities beginning a water conservation program.

- Most of the progress in laws and product testing has originated from AWWA's Water Conservation Division, a voluntary organization with members throughout North America. However, California has long been a leader in water conservation programs, developing new technologies and approaches that have been copied nationwide (and around the world). In 1991, water providers throughout California signed a Memorandum of Understanding (MOU) pledging to carry out cost-effective water conservation Best Management Practices. The California Urban Water Conservation Council (CUWCC), a nonprofit organization that is the "keeper" of the MOU, developed research to support implementation of these Best Management Practices (BMPs) (www.cuwcc.com/).

- The 1992 Energy Policy Act (EPAct) requires that maximum toilet flush volumes be no more than 1.6 gallons, starting January 1, 1994. EPAct also set maximum flow standards for showerheads, faucets, and urinals.
- In 1998 the US Environmental Protection Agency (USEPA) issued guidelines for water conservation plans for public water systems, www.epa.gov/watersense/pubs/guide.htm.
- In 1999, the American Water Works Association Research Foundation (currently the Water Research Foundation) published *Residential End Uses of Water Study*, documenting a survey of metered water use in 12,000 residences and measured end uses in 1,188 homes in 12 cities across the United States and Canada (Mayer et al. 1999). This provided valuable benchmarking data.
- In 2001, a *Handbook of Water Use and Conservation: Homes, Landscapes, Businesses, Industries, Farms* by Amy Vickers (WaterPlow Press) first provided an up-to-date guidebook for utilities of all sizes.
- In 2006, AWWA published M52 *Water Conservation Programs—A Planning Manual*, first edition, by William O. Maddaus and Lisa A. Maddaus.
- In 2003, the Maximum Performance (MaP) testing program for toilets was funded by 22 U.S. and Canadian water agencies, to identify how different toilet models perform, and certify and rank high performing models. This testing protocol serves as the basis for the USEPA WaterSense label for high-efficiency toilets (Figure F-1).

The nonprofit Alliance for Water Efficiency (AWE) was formed in 2006, following a series of focus groups on its mission and structure. The AWE is based in Chicago and is supported by member utilities, foundations, and private sector companies. The AWE website is an increasingly rich source of up-to-date information, www.allianceforwaterefficiency.org.

Figure F-1. MaP testing of toilet performance, Veritec Consulting facility near Toronto. (Photo: Water Media Services)

- In 2006, the USEPA launched its WaterSense program. Following the success of the USEPA's well-known Energy Star program, WaterSense aims to "enhance the market for water-efficient products and services by building a national brand for water efficiency." WaterSense certifies high-efficiency toilets, faucets, irrigation professionals, and continues to develop protocols and contract for third party testing of other products. An emphasis of WaterSense is to promote products that use at least 20 percent less than required by the 1992 EPAct and to contribute to a market transformation to these more efficient products.
- The Smart Water Application Technologies, or SWAT program, is a national partnership initiative of water purveyors and irrigation industry representatives through the Irrigation Association, www.irrigation.org. In the same way that the performance testing of toilets has propelled improvements, development of testing protocols and testing of irrigation equipment is stimulating improvements for irrigation equipment.
- The Awwa Research Foundation (currently the Water Research Foundation) in 2007 published *Water Efficiency Programs for Integrated Water Management,* which summarizes water use efficiency savings and cost assumptions.
- By the time you are using this guidebook, considerable changes in laws and technologies may have already occurred in this evolving field. AWWA maintains the WaterWiser website (www.waterwiser.org) as a source of up-to-date information.
- The AWWA Water Library allows online access by members to earlier publications. apps.awwa.org/WaterLibrary/Search.aspx
- A monthly online newsletter *Water Efficiency* News is published by the Alliance for Water Efficiency and is available on its web site at www.allianceforwaterefficiency.org.

Introduction

SUMMARY:

- *Defines water conservation*
- *Defines water use efficiency, demand management and water conservation measures*
- *Distinguishes between conservation and curtailment*
- *Distinguishes between long-term conservation and short-term emergency management*
- *Discusses reasons for conserving, from a variety of perspectives*

WHAT IS WATER CONSERVATION?

Growing population, environmental concern, climate change, periodic droughts, and economic considerations all indicate the need for wise stewardship of finite freshwater resources. Water conservation is one tool that can be used effectively to meet challenging constraints and opportunities.

The reasons for conserving water and the ways to accomplish conservation will differ for water utilities, depending on geography, number of customers, rate of population growth, growth in water demand, and other factors. Not all options are appropriate for all situations, and utilities can choose the type and extent of measures that best fit their requirements.

Water conservation can be described as "any beneficial reduction in water use or in water losses" (American Public Works Association 1981). Ways to achieve water conservation include

- Technical measures (such as water-saving fixtures, efficient irrigation systems, and reduction of water losses in the system)
- Process or use changes (such as more efficient industrial processes, water reuse, and low water-use landscaping)
- Regulatory action
- Metering and pricing policies

- Public education to help people change behaviors to use less water without impacting life styles.

Water efficiency or water use efficiency refers to the accomplishment of a function, task, process, or result with the minimal amount of water feasible. It is also an indicator of the relationships between the amount of water needed for a specific purpose and the amount of water used, occupied or delivered (Vickers 2001). Water efficiency is a tool of water conservation that reduces water demand without changing the quality of the use.

The term *demand management* allows a contrast with supply-side management and source development. Demand management emphasizes the quantifying of results at the level of master meter reads for system-wide flows or at the customer billing level.

Conservation should be distinguished from *curtailment*, which means mandatory reduction in water use. Curtailment is only necessary during drought or emergency situations and requires measures that can achieve immediate results. Water conservation in contrast allows reduction in use without changing the level of customer service.

A water conservation *measure* is an action, behavioral change, device, technology, or improved design or process implemented to reduce water loss, waste, or use. Note that the value and cost-effectiveness of a water efficiency measure also involves its effects on the use and cost of other natural resources (for example, energy and/or chemicals) (Vickers 2001).

Conservation can be used to reduce both annual average demand and monthly, daily, or hourly peak demand. Peak and average demand have different effects on the need for capital facilities. Reducing average demand principally affects raw water storage requirements. Reducing peak demand reduces the costs for new treatment, conveyance, and distribution and can save millions of dollars by delaying or eliminating the need for additional reservoirs, wells, or treatment facilities.

Changes in indoor plumbing fixtures can reduce average demand, but will have little effect on summer peak demand. However, implementing seasonal pricing and low water-use landscaping can help reduce summer peak demand.

WHY SHOULD UTILITIES EMPHASIZE WATER CONSERVATION?

The utilities that have put the most effort into water conservation to date are in arid regions and other areas where there are severe supply limitations. Benjamin Franklin's statement, "When the well runs dry, we know the worth of water" has unfortunately been proven true. Out of necessity, these utilities have worked toward technological advances, often accompanied by regulatory (legislative) mandates.

As a current example of water conservation through necessity, the San Antonio Water System (SAWS) must buy additional water at prohibitive costs from a neighboring utility with prior rights. The utility has found its investment in toilet replacements and other conservation retrofits to be the most cost-effective way to produce "new" water. The City of Santa Fe, New Mexico, had developers seeking permits but no additional water capacity, so these developers had to "find" water by retrofitting older homes and buildings. The term *capacity buy back* has been coined for this process and is being widely used in Canada (Gauley 2002).

In fact, the results of these forced efforts demonstrate that water conservation is a cost-effective means of providing additional water. "The cheapest water you will ever find is the water you already have in your system," is a maxim of the water conservation community.

Water Utility Perspective

Decreasing demand and increasing system-operating efficiencies can have numerous benefits to utilities, including

- Decreasing reliance on development of new sources to meet demand, important because high-quality surface and groundwater supplies are becoming harder to find at reasonable costs. Debt service for new sources can be postponed.
- Downsizing or delaying capital facilities for new water supply, transmission, storage, and treatment.
- Reducing operation and maintenance costs that depend on demand (pumping and chemical costs, for example), providing public health and economic benefits.
- Reducing energy costs of operation and reducing resulting carbon emissions.
- Addressing community values and expectations by developing and implementing conservation measures that make sense for the local area. The general public is becoming increasingly aware of the need to use natural resources wisely.
- Improving supply reliability, which can reduce frequency and duration of water use curtailment in droughts.
- Providing confidence to the public by improving their perception that the utility is taking all possible steps before major expansions and incurring more costs.
- Demonstrating water-use efficiency to regulatory agencies.
- Relying on conservation measures instead of developing new supplies may also have a few disadvantages for the utility if not properly addressed in advance:
- Conserving water can reduce revenues through reduced water sales, unless sufficient advance planning has been done and rate structures revised to be revenue neutral. When designing future rates, utilities should take into account projected lower revenues, the costs of the conservation program, and cost savings resulting from lower operating costs and deferred capital facilities. Information on customer response to rate changes and other conservation measures, based mostly on income level, is available for some areas and can guide a utility in designing rates that maintain stable revenues and assign proportionate costs to high water users (Chapter 3, Pricing).
- In some locations, especially in the western United States, conservation could threaten the "use it or lose it" doctrine of water law and water rights.
- Conservation programs, particularly rebate programs, can be expensive and should be designed carefully to be cost effective for the individual utility.
- Short-term drought savings may be more difficult to achieve and the amount of water that can be saved by water rationing reduced, if water conservation reduces the "slack" in the system. However, if the conservation ethic is a part of

the community, in part due to utility conservation programs, customers seem to be more receptive to requests to reduce water use.

- Public works staff may express concern that reducing indoor water use may lead to increased sewer system maintenance. Such problems have not been documented, however, probably because sewer velocities remain essentially the same at lower flows. A 2009 Canadian study is addressing this concern.

Wastewater Utility Perspective

Lower residential water use and decreased industrial/commercial consumption will result in reduced wastewater flows. This helps wastewater utilities by saving energy (i.e., pumping costs), decreasing the amount of treatment chemicals, and possibly reducing capital costs for additional treatment and collection facilities. These savings are typically realized for systems with separate stormwater and sewage collection facilities.

Stormwater Management Perspective

As homes have been built more densely, water recharge areas have decreased or have been eliminated, and flooding problems have increased. Outdoor water conservation programs can decrease stormwater runoff problems.

When sprinklers are adjusted to avoid spraying onto streets and other hard surfaces, and adjusted to avoid puddling and over-watering, nonpoint source pollution is reduced. Run-off loaded with herbicides and pesticides traveling through stormwater systems to streams and rivers can be lessened. Where rains occur during the warm irrigation season, use of rain sensor devices stops irrigation runoff from adding to the stormwater stream during rain events.

Rainwater harvesting, where not prohibited by local codes, helps stormwater management by utilizing rainwater on-site.

Customer Perspective

Water conservation benefits the customer through lower annual water and sewer costs. Customers may also experience lower energy bills because of decreased use of energy needed to heat water for showers, faucets, dishwashers, and clothes washers. Septic tanks may function better without excessive flows. Industrial customers may have lower pretreatment costs as a result of reduced water use.

Concerns about the environment may also be important to many customers. There is a growing interest in *sustainability* to ensure the needs of existing and future generations are met and that habitats and ecosystems are protected (US Environmental Protection Agency 1992).

Legislative Mandates

Many utilities are establishing water conservation programs in response to federal, state, or local regulatory requirements. In some cases, regulatory agencies are requiring utilities to institute water conservation programs before authorizing the development of new water sources, renewing permits for existing sources, or the construction of capital facilities.

Federal agencies encourage conservation by sponsoring programs, developing guidelines, and linking water conservation programs to federal grants for water and wastewater facilities. To further support of water conservation, federal actions have changed plumbing and appliance codes by requiring the use of low-flow plumbing fixtures and water-efficient appliances such as dishwashers and clothes washers. (Maddaus et al. 2001).

Federal agencies that are active in water conservation are the US Environmental Protection Agency (USEPA), US Army Corps of Engineers (USACE), US Department of Interior (DOI), Bureau of Reclamation (USBR), the Department of Housing and Urban Development (HUD), and the Federal Energy Management Program (FEMP), Water Efficiency section. The US Geological Survey (USGS) collects and maintains water use data. Appendix A lists federal and state agencies with regulatory authority or other relation to the field of water conservation.

Environmental Perspective

Water conservation can make more water available for environmental uses, such as river floodplains and wetland wildlife habitats, and in-stream flows for fish. A conservation program accommodates the concerns of environmental groups by demonstrating that the utility has explored all available options before developing a new water source. A utility that fails to make this effort faces almost certain opposition to its decisions by environmental groups, as well as by regulators and utility customers. Conservation allows the utility to demonstrate environmental awareness and responsibility.

HOW TO BEGIN

A small or medium-sized utility should designate one "point person" as Water Conservation Coordinator. This person will likely have other related duties within the utility. A team should be designated within the utility to plan and work with this coordinator to implement the conservation program. More information on staffing and coordination is provided in Chapter 6.

The greater the available budget for a water conservation program, the more that can be accomplished. An absolute minimum budget, for example $1,000, should be set aside for the program. A major goal in the beginning months or years of a program is to establish confidence in the program from the utility and/or municipal administration in order to have additional funds budgeted each year according to need. Chapter 3, Funding, discusses possible funding sources aside from the general utility budget.

Key partners will be the utility's finance, public affairs, and customer service departments. Finance will be able to provide information to assist in preparing the utility water use profile (Chapter 2). Access to billing data will assist later in documenting water use before and after water conservation measures at a customer level. The public affairs and customer service departments will also assist in outreach to water customers. At the same time, the water conservation program will benefit customer services in providing positive responses to customer complaints.

If the motivation to begin a water conservation program is regulatory, the local regulatory agency should be consulted for specific requirements. Find out what assistance these agencies provide for small and medium-sized utilities. Assistance may include cost-share funding for water conservation measures that have shown quantitative demand reductions. Chapter 4 reviews some of the measures that have proved successful around the US that can be used to quantitatively reduce demand.

If the motivation to begin the water conservation program is for public approval, the utility should focus initially on a strong public information component. Chapter 3 gives tips on approaches to public information, and Appendix C provides some sample materials. Because public information does not always translate into changed customer behavior, many utilities reserve their strongest public information efforts for drought periods.

Overall, a strong public information program is essential for creating a strong community response toward water conservation. The utility should retrofit its own facilities and demonstrate a good example to the public.

If the motivation to begin the water conservation program is finding a more economical alternative to costly new water supplies, a more detailed analysis than is possible in this guidebook will be required. Chapter 6 will suggest sources of information to make these cost-effectiveness comparisons among different alternatives.

This guidebook identifies references to obtain more technical information or pursue more detailed methodologies. However, for the most part, a utility will not need to consult other documents or hire specialists to establish a conservation program.

The worksheets in Chapter 2 produce a utility water use profile and are a good baseline of information that can be carried forward as the program develops. A more detailed approach is presented in AWWA's M52 *Water Conservation Programs—A Planning Manual.*

Based on progress previously made in the water conservation field, including technological advances, better data on savings, and better documentation methodologies, establishing a water conservation program is currently easier than in the past. Even if the program cannot fully reduce the demand to the degree required, a utility can "chip away" at the demand, and with the opportunity to prove itself and with increased funding, can accomplish far more.

Utility Evaluation

SUMMARY:

- *Shows how to determine the utility's current water usage*
- *Shows how to characterize the utility to identify best conservation opportunities*

OVERVIEW OF PLANNING

This chapter provides guidance for the organization of information about a water system in order to target the best opportunities for water conservation. With knowledge of a utility's supply and demand characteristics, community ages, and use by different sectors, goals can be set for the program and the most promising and cost-effective conservation measures can be selected.

Detailed planning methods will allow determination of future water supply needed by a utility in regards to anticipated future growth. For methods accepted by specific states, and methods to determine what proportion of future supply needs to come from water conservation, the local state regulatory agency (Appendix A) should be consulted. The AWWA M52 *Water Conservation Programs—A Planning Manual* shows how to determine the benefit-cost ratio of water conservation measures in comparison to other water supply options of a utility.

While determining the comparative value of water conservation and alternative water supplies for a utility is beyond the scope of this guidebook, some information on a utility's distribution level (supply side) is important to target a water conservation program, in addition to information on customer level (demand side).

For purposes of starting a program, information on peak demand is necessary. The following text and Worksheet 1 relate to obtaining that information. The utility will also need to compile information on specific high water users, on the age of housing stock, and on customer classes in WORKSHEETS 2A-2D. Information compiled in these worksheets will be used in Chapter 6, which illustrates how to design a conservation program.

Replacement of older inefficient fixtures with newer high efficient fixtures will represent the most dependable means of reducing water and wastewater use. Measures that reduce peak demand through outdoor water use management can produce very effective conservation savings with high benefit-cost ratios.

FACTORS DETERMINING FUTURE WATER DEMAND

A utility will most likely have created a water supply master plan for internal financial reasons or as required by state regulations.

Future demand typically depends on

- Land use changes
- Resident and seasonal population
- Number, value, and type of housing units
- Landscaped area
- Landscaping practices
- Employment
- Water and wastewater pricing
- Agricultural use
- Personal income
- Climate and weather conditions
- Conservation activities
- Environmental constraints

DETERMINE CURRENT WATER USAGE

Step 1: Describe the Service Area

Using WORKSHEET 1, at the end of the chapter, gather as much of the following information about the service area as possible.

- **Current population.** Use US Census Bureau or state (possibly university) estimates.
- **Future population.** Check with state or local government agencies and regional planning groups. Typical population growth projections will span 5-, 10-, or 20-year periods.
- **Number of connections.** Utilities usually have this data broken down by type of account.

A complete water supply master plan will contain much of this information.

Step 2: Describe Water Use

Add to WORKSHEET 1 as much of the following information as possible:

- **Average annual water use,** in millions of gallons per year (mgy) or in acre-feet per year. To calculate this number, add the total water use for the past 5 years and divide by five. Exclude drought years when water use restrictions were in effect because water usage during those years will not be typical. Shorten the averaging period to 3 years if growth has been rapid (for example higher than 3 percent per year).

- **Nonrevenue water.** Formerly, a typical (conservative) figure for "unaccounted-for water" in Western urban areas was 10 percent of total water production, and older urban systems (typical in the East) often see 20 percent unaccounted-for water. A new method has been developed to account for all system water produced, and a more detailed analysis of non-revenue water according to this methodology is now replacing estimates of unaccounted-for water (See Chapter 3).
- **Average daily use.** Divide average annual use by 365.
- **Average daily summer use.** Using water production records, divide water use during June, July, and August by the number of days in these months.
- **Peak day use.** Identify the peak-use day from water production records. The peak day in much of the US occurs in the warm months of July or August. In states with a summer rainy season, the peak day is generally the warmest day prior to onset of the rainy season. For example, in Florida this peak day typically occurs in May. In part of the West, the peak day may be in September.
- **Peak day to average day use ratio.** Divide the peak day use by the average daily use.
- **Estimated seasonal use.** Nonseasonal uses are assumed to be constant over the year and include indoor water usage from appliances, plumbing fixtures, and other items that are not dependent on weather. Seasonal uses are typically associated with lawn watering and peak during months of maximum temperature and minimum rainfall.

Use the following steps for determining seasonal use.

1. Identify the month with the lowest water demand (usually this is January). Demand occurring in this month is defined as the nonseasonal use. Seasonal use is defined as zero in this month.
2. Identify the average demand in this month.
3. Multiply the average demand in the lowest month by 12 to get nonseasonal annual water use.
4. Divide the nonseasonal annual water use by average annual water use. Multiply by 100 to get a percentage.
5. Subtract the nonseasonal percentage from 100 to determine the seasonal annual use percentage.

$$\text{Seasonal Use, \%} = (1 - \text{low month} \times 12/\text{Annual Use}) \times 100$$

CHARACTERIZE YOUR SERVICE POPULATION

Step 1: Identify highest use customers

High water-use customers are excellent targets for cost-effective water conservation programs. It should be noted whether the use by each high use customer correlates with utility summer peak demand or if it is relatively constant year round.

WORKSHEET 2A takes information from a simple demand report that most utility billing systems can prepare. The information on the worksheet should be recorded to refer to and to share with the team planning the water conservation program or copies of demand reports can be compiled and annotated prior to this discussion.

Step 2: Identify age of residential accounts

Replacement of older inefficient fixtures with newer high efficiency fixtures represents a large portion of potential water/wastewater savings (Maddaus 2004). WORKSHEET 2B is designed to identify opportunities for plumbing retrofits in older homes.

For numbers of households in each period, use the US Census website (http://factfinder.census.gov/) or university generated population data, possibly accumulated in a utility water supply plan. If this source does not indicate actual number of homes, the data source should have an average number of residents per household, which when divided into total population will allow an estimated number of homes. A utility may also have historic data on number of accounts enrolled in different time periods.

For toilet rebates the actual years of interest are 1985, before which 5-7 gallon per flush toilets were sold and after which 3.5 gallon per flush toilets were sold, and 1994, after which 1.6 gallon per flush toilets or less were sold. Unfortunately, these years of interest in regard to national fixture requirements do not correlate perfectly with census data that is by decades. Approximations will have to be made. It should be noted that there is a natural rate of replacement of fixtures, because of leaks, breakage, or remodeling. This natural replacement rate has been estimated by the California Urban Water Conservation Council to be about 4 percent in California. Replacement of water-using household fixtures to current more efficient models above that 4 percent natural replacement rate will be an important effort, as outlined in Chapter 4.

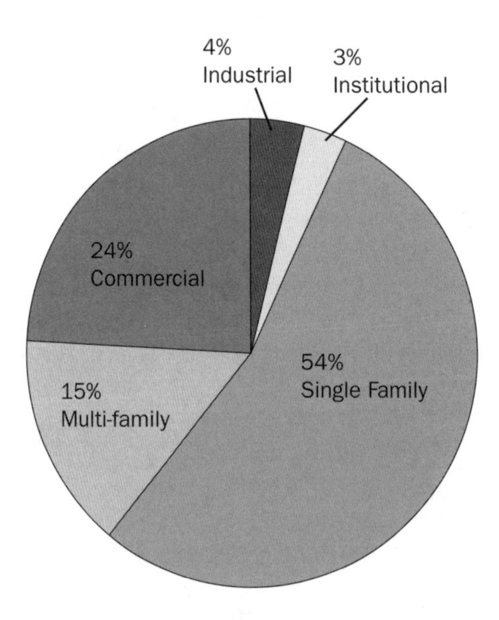

Figure 2-1. Proportion of total U.S. water sales going to different sectors. Single family makes up 54 percent. (from Maddaus et al. AWWA M52 *Water Conservation Programs—A Planning Manual* 2006)

Step 3: Identify high irrigation use accounts

WORKSHEET 2C is designed to assist a utility in collecting information on high irrigation users. If a utility does not show an increase in summer peak demand, irrigation may not be the most important water conservation target and working with this sector may be unnecessary. However, if accelerated home building is anticipated, it should be noted that builders in most parts of the US currently install landscapes with automatic irrigation systems, and measures to reduce residential outdoor water use (Chapter 4) will likely be cost-effective.

Step 4: Identify commercial, institutional, and industrial accounts

Use WORKSHEET 2D to compile information on the different sectors within Commercial, Institutional, and Industrial (CII) areas. Although CII accounts make up the minority of accounts in most utility service areas, they may use a large percentage of a utility's water and are therefore great candidates for successful demand reduction. This information will allow a utility to target programs proven successful elsewhere, as outlined in Chapter 4.

A utility's billing system may have accounts divided into Single Family Residential, Multi-family Residential, Commercial, Institutional, and Industrial, or it may be only divided by meter sizes. The most advanced billing systems identify particular businesses by North American Industry Classification System (NAICS) codes, formerly called SIC codes. Codes that allow sorting of accounts by actual type of business make this kind of analysis very easy. If such reports are not available, staff members' knowledge of the service area may be able to generate this information with reference to the Chamber of Commerce or "Yellow Pages."

Worksheet 1
Describe the Service Area
Provide the following information as best as possible.

- **Current population** _______________
- **Estimated future population in**
  - 5 years _______________
  - 10 years _______________
  - 20 years _______________
- **Extent of metering** (percent)
  - Residential
  - Single-family _______________ %
  - Multifamily _______________ %
  - Commercial _______________ %
  - Institutional _______________ %
  - Industrial _______________ mgd
- **Number of residential service connections:**
  - Single-Family _______________
  - Multifamily _______________
- **Number of nonresidential service connections:**
  If the billing system does not divide as below and this information is not otherwise available, change the categories to meter sizes (¾ in., etc.) and fill in:
  - Commercial _______________
  - Industrial _______________
  - Institutional _______________
- **Usage from unmetered supply** (if unmetered supply, see Appendix E for method) _______________ mgd
- **Total nonrevenue water** (see chapter 3) _______________ mgd
- **Average daily use** _______________ mgd
- **Average daily summer use** (over last 5 years):
  - Average total water use for June/July/August _______________ mgd
  - Number of days in these months _______________ days
  - Average daily summer use _______________ mgd
- **Peak day use** _______________ mgd
- **Peak day to average day use ratio** _______________
- **Estimated seasonal use:** _______________
  - Month with lowest water demand _______________
  - Average demand in this month _______________ mgd
  - Nonseasonal annual water use _______________ mgd
  - Nonseasonal use, percent of total _______________ %
  - Seasonal annual use, percent of total _______________ %
- **System safe yield** _______________ mgd
- **Supply or permit limitations** _______________ mgd

Worksheet 2A
Identifying Water Conservation Opportunities: Top 10 Water Users

The utility can provide an annotated demand report instead of filling in this worksheet. Identifying the top twenty or 100 users is more valuable than top 10, but this is a starting point. The top water users may include all customer classes or do this separately for SF, MF, Commercial, Institutional and Industrial, if customer class data is available. If no customer class data is available, separate by meter size. Note whether use has a distinct seasonal peak.

Date: ______________________

Customer name	Sector/Peak use?
1.	
2.	
3.	
4.	
5.	
6.	
7.	
8.	
9.	
10.	

Date: ______________________

Customer name	Sector/Peak use?
1.	
2.	
3.	
4.	
5.	
6.	
7.	
8.	
9.	
10.	

Worksheet 2B

Identifying Water Conservation Opportunities: Classifying customers by year of construction and fixture installation

(for plumbing retrofit opportunities)

Residential Accounts by Year of Construction, focusing on plumbing retrofits

Single-family residential (SFR)

- Total existing SFR accounts ______________
- SFR accounts in homes built before 1980 ______________
- SFR accounts in homes built from 1980-1989 ______________
- SFR accounts in homes built from 1990-1999 ______________
- SFR accounts in homes built from 2000-2009 ______________
- SFR accounts in homes built from ___________ ______________
- SFR not yet built but permitted ______________

Multi-family residential (MFR), master metered

- MFR accounts built before 1980 ______________
- MFR accounts built from 1980-1989 ______________
- MFR accounts built from 1990-1999 ______________
- MFR accounts built from 2000-2009 ______________
- MFR accounts built from ___________ ______________
- MFR not yet built but permitted ______________

Multi-family residential, submetered

- MFR accounts built before 1980 ______________
- MFR accounts built from 1980-1989 ______________
- MFR accounts built from 1990-1999 ______________
- MFR accounts built from 2000-2009 ______________
- MFR accounts built from ___________ ______________
- MFR not yet built but permitted ______________

Worksheet 2C
Identifying Water Conservation Opportunities: Residential Accounts, focusing on irrigation use

Category	**Number**
Total Single-family Residential (SFR) Accounts	________
SFR customers with separate irrigation accounts	________
SFR customers with reclaimed water irrigation account, unmetered	________
SFR customers with reclaimed water irrigation account, metered	________
SFR customers with private irrigation sources-wells, lakes (estimate)	________
Condominiums and apartments may be classified as multi-family residential by the utilities. Fill in notes on what type of Multi-family Residential as this information is compiled.	
Multi-family residential, with separate irrigation accounts	________
MFR accounts with reclaimed water irrigation account, unmetered	________
MFR customers with reclaimed water irrigation account, metered	________
MFR customers with private irrigation sources-wells, lakes (estimate)	________
Residential Accounts, focusing on quantity of use	
SFR customers using over 10,000 gpm	________
SFR customers using over 30,000 gpm	________
SFR customers with separate irrigation accounts, using over 30,000 gpm	________
SFR customers with reclaimed water irrigation accounts, metered, using over 30,000 gpm	________
MFR customers using over 10,000 gpm	________
MFR customers using over 30,000 gpm	________
MFR customers with separate irrigation accounts, using over 30,000 gpm	________
MFR customers with reclaimed water irrigation accounts, metered, using over 30,000 gpm	________

Residential areas (subdivisions) where irrigation use is high and presents peak demand difficulties for the utility.

Name of subdivision	**HOA present**
1.	
2.	
3.	
4.	
5.	
6.	
7.	
8.	
9.	
10.	

Worksheet 2D

Identifying Water Conservation Opportunities: Commercial, Institutional, and Industrial Accounts

Note that this worksheet will take some time to fill out, depending on the detail on account identification in the billing system. Lists of schools and restaurants are best obtained from the phone book (which includes online versions) or perhaps chamber of commerce information.

Commercial and Institutional Account Estimates

Category	#Accts (approx)
Hotels or time-shares	
Assisted Living Centers	
Mobile Home Parks	
Gyms	
Colleges, with residences	
Colleges, commuter	
Elementary Schools	
Middle Schools	
High Schools	
Colleges	
Restaurants, sit down	
Restaurants, fast food	
Prisons	
Hospitals	
Municipal buildings	
Parks with irrigation	
Condominiums	
Apartments	
TOTAL	

NOTE: Condominiums and apartments may be classified as multi-family residential by some utilities.

Review Utility (Supply Side) Conservation Measures

SUMMARY:

- *Describes potential conservation measures that can be performed by the utility.*

The following measures are described in this chapter:

- System audit
- Utility leak detection and repair
- Pressure management
- Metering
- Pricing
- Funding the water conservation program
- Staffing
- Public Information

SYSTEM AUDIT

The amount of water lost by the 55,000 community water systems in the U.S. is around 6 billion gallons per day, or enough to meet the delivery needs of the country's 10 largest cities (Thornton et al. 2008). Most utilities have some amount of water produced but not sold or otherwise accounted for. This has commonly been called *unaccounted-for water* (UAW), and regulators have required this to be under a certain percentage, for example under 10 percent. Bond ratings for utilities may also require UAW under 10 percent.

However, with metering becoming more universal, regardless of a water system's size, water loss should be expressed in terms of actual volume, not a percentage. In this way, the monetary value of water loss can be estimated, according to the AWWA M36 *Water Audits and Loss Control Programs.*

American Water Works Association and the International Water Association (IWA) jointly developed a reliable water audit methodology in 2000. State and regional regulatory agencies in the United States now embrace this AWWA/IWA Water Audit Methodology as 1) an improved and reliable practice compared to the imprecise unaccounted-for water process, and 2) a standardized approach that can produce data which allows performance comparisons and benchmarking of best practices.

Water Audit Software, developed as part of the new IWA-AWWA methodology by AWWA's Water Loss Control Committee and accompanying the M36 manual, is Excel-spreadsheet based. Released in continually updated versions, the software is available free from the AWWA WaterWiser website, www.waterwiser.org/. The AWWA Manual M36 *Water Audits & Loss Control Programs*, 2009, can be purchased from the AWWA bookstore. Visit www.awwa.org or call 800-926-7337.

With these new tools, the water industry is undergoing a paradigm shift toward sound water loss management. In this approach, utilities provide accountability in their operations by auditing their supplies, and implementing controls to keep system losses to reasonable minimal levels. With the new methodology, unaccounted-for-water is becoming a term of the past. The focus has turned to *nonrevenue water* (NRW).

Real loss, apparent losses and unbilled authorized consumption make up NRW, according to the IWA water balance methodology (Figure 3-1). Real losses are water lost from the distribution system through leaking pipes, joints, and fittings; leakage from reservoirs and tanks; reservoir overflows; and improperly opened drains or system blow-offs. Apparent losses are inaccuracies in customer metering, consumption data handling errors, or any form of theft or illegal use. All of these cost utilities needed revenue.

The unbilled authorized consumption portion of NRW includes water supplied for the utility's operations and the operations of the associated municipality. Even if unbilled, this water use should be metered and accounted for. The IWA-AWWA methodology is most successful with full metering. For example, the San Francisco Public Utilities Commission Water System has had full metering of ALL accounts regardless of billing status. Municipal customers, such as the city's own parks and public buildings, are metered although not billed. Even if not all customers are billed, metering allows the consumption to be entered into the NRW category, revealing possibilities for better management and revenue recovery. California's Proposition 218 requires utilities to bill all accounts, even those within a utility's own municipal government.

Several states (including Texas, Washington, and California), now require utilities to use the IWA–AWWA methodology. Texas, which was the first state to enact a requirement, has a large number of small utility districts, and standardization of water loss data collection even in these small systems is a positive step forward.

Inefficiencies in water distribution systems, including underground water system leakage, result in loss of revenue and less water available for purchase by retail customers. Methods to detect, locate, and correct leaks exist and are improving.

There are two steps to measure NRW. First conduct a water system audit, which is relatively easy to perform for metered systems. Appendix E includes a method for unmetered systems. The estimated time required to perform a first-time audit ranges from 2 weeks of one technician's time for a small retail water agency to 26 weeks for a large agency. Subsequent audits should require about half this time, because procedures will be in place.

System input volume mgy	Authorized consumption mgy	Billed authorized consumption mgy	Billed metered consumption (including water exported) Unbilled metered consumption (including water exported)	Revenue water mgy
		Unbilled authorized consumption	Unbilled metered consumption	Nonrevenue water mgy
	Water losses mgy	Apparent losses mgy	Unauthorized consumption Customer metering inaccuracies Data handling error	
		Real losses mgy	Leakage on Transmission and/or distribution mains Leakage and overflows at utility's storage tanks Leakage up to point of customer metering	

Figure 3-1. Components of water balance for a transmission or distribution system. A utility can focus on nonrevenue water to find opportunities to recoup revenues at a system level. (Maddaus et. al 2006)

Then compare the amount of water entering the distribution system to the amount of water supplied to customers. The accuracy of all agency meters and a random sample of customer meters should be checked.

Nonrevenue water, including both unbilled authorized consumption and water losses, can be expressed in millions of gallons per year (mgy) or acre-feet per year or in dollars of revenue lost.

The use of computer software such as the AWWA Water Loss software can reduce the time required to analyze audit results. This effort is justified if the difference between the source water meter readings and water sales is greater than 10 percent.

UTILITY LEAK DETECTION AND REPAIR

System leak detection and repair can be a cost-effective water conservation measure to reduce NRW. It should be noted that leak detection and repair achieves year-round conservation, as opposed to just peak-use reduction.

If the audit indicates significant water losses from leakage, a utility may decide to conduct a leak detection and repair program. Leaks occur because pipes corrode with age. The extent of corrosion depends on water chemistry and surrounding soil chemistry. Leaks are also caused by settlement, which can cause pipes to crack or joints to separate. As shown in Figure 3-2, background leakage is generally unreported and undetectable using traditional acoustic equipment. Unreported leakage often does not surface but is detectable using acoustic equipment, as in Figure 3-3. Reported leakage is often at the surface and reported by the public or utility workers.

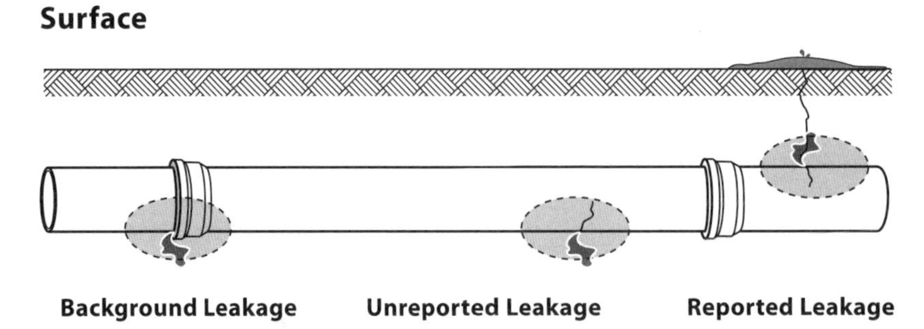

Figure 3-2. Types of real losses in water pipes. Background leakage is unreported and undetectable using traditional acoustic equipment. Unreported leakage is detectable using acoustic equipment. Reported leakage at surface is reported by public or utility workers. (Graphic: Tardelli 2005 in Thornton et al. 2008, Water Media Services)

Figure 3-3. Leak detection with acoustic equipment (Photo: Miami-Dade Water and Sewer)

Some state agencies loan out leak detection equipment and offer staff training. The utility should check with the USDA Rural Development (formerly known as Farmers Home Administration) and state agencies listed in Appendix A for possible grants or loans for leak reduction. The National Rural Water Association and its state affiliates may also provide assistance.

Meter calibration, valve maintenance, and corrosion control can result in additional water savings. These actions may also result in secondary benefits such as reducing wastewater flows due to infiltration, reducing the utility's liability for property damage in the event of a major distribution system leak, and correcting under-registration of meters for large customers.

PRESSURE MANAGEMENT

Reducing water pressure reduces stress on pipes and joints. Lower water pressure may slow system deterioration, lessening the need for repairs and extending facility life. It can also help reduce wear on end-use fixtures and appliances.

Pressure management and reduction strategies must be consistent with state and local regulations and standards, as well as take into account system conditions and needs. Obviously, reductions in pressure should not compromise the integrity of the water system or service quality for customers.

For residential areas, pressures above 80 psi should be assessed for reduction. A more aggressive plan may include the purchase and installation of pressure-reducing valves in street mains, as well as individual buildings. Utilities might also insert flow restrictors on services at the meter. Restrictors can be sized to allow for service length, system pressure, and site elevation. Utilities can consider providing technical assistance to customers, particularly large-use customers, to address pressure problems and installing pressure-reducing valves to lower the customers' water pressure (EPA 1998).

METERING

In the early 1980s, Denver Water, the provider to the city of Denver, Colorado, found that installation of meters resulted in water savings averaging about 20 percent (Maddaus 1987). These savings resulted from conversion from a flat-rate type of billing to billing based on metered consumption. Metering without a volumetric (use-based) water rate is unlikely to produce savings. Denver's savings ranged from 2 percent in winter to 25 percent in summer (due to reductions in landscape irrigation). Humid areas that require less outdoor irrigation may realize lower savings than 20 percent, while drier areas could realize higher savings. The California Urban Water Conservation Council in 2003 referred to metering as reducing demand by an average by 20 percent (Chesnutt et al. 2007).

Labor costs are the biggest factor in meter retrofit programs and account for the wide cost variation observed in retrofits. Manual versus machine excavation and the cost of landscape and driveway replacement affect cost.

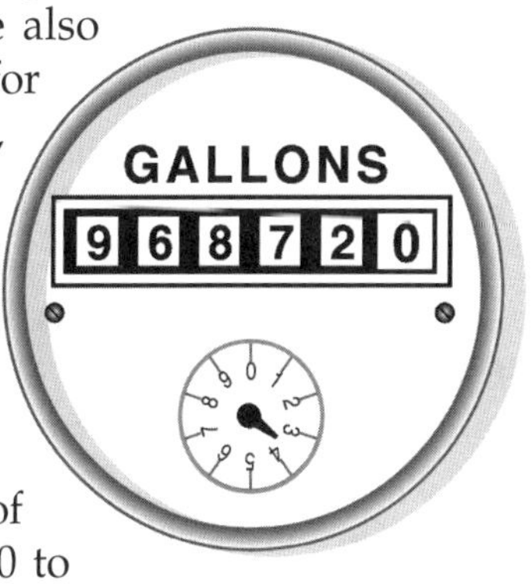

Reclaimed water should also be metered. Many utilities are using treated wastewater effluent in place of potable water for irrigation by large customers, like golf courses, schools, and businesses. Some are also supplying reclaimed water to residential customers for irrigation. If these connections are metered from the start, the utility can charge volumetric rates. For example, by the year 2000, the City of Port Orange, Florida, provided reclaimed water to approximately 10 percent of its 20,000 residential service connections; however, this reclaimed water was unmetered and sold at significantly lower rates than potable water. During the 1998-2001 Florida drought, customers overused this inexpensive water to the point of danger to system pressures. The city then spent $750,000 to meter all reclaimed customers. Making this effort allowed this medium-sized utility to charge volumetric rates, which has assisted in irrigation demand reduction.

Submetering is the addition of separate meters to indicate individual water use in apartments, condominiums, and mobile homes. Residents of individual units conserve when they are billed according to actual use. Most utilities have chosen not to submeter because of costs of meter installation and additional meter reading and billing work for the utility. However, a National Multiple Family Submetering and Allocation Billing Program Study found that submetering on multifamily apartment units, with billing based on actual consumption, results in water savings of 15 percent (Aquacraft 2004). Other studies demonstrated reduced consumption up to 39 percent through submetering (Chesnutt et al. 2007).

Throughout this guidebook, metering will be seen as a key element. Another tried and true water conservation maxim is "You can't manage what you don't measure."

PRICING

To provide water service to its customers, a water agency must receive sufficient revenues to recover its costs, including operation and maintenance costs, capacity costs, customer costs, and administrative costs (AWWA 2000). Maintenance of utility revenues and conservation might seem to be mutually exclusive goals, but they can be very compatible when a utility is short of supply. Because high water-use customers contribute significantly to utility revenues, if these users cut their demand, revenues will certainly drop. However, careful planning can ensure revenue stability to the utility when that happens. In a utility environment in which large contingency fund reserves are difficult to accumulate, a utility must be able to accurately predict what will happen when rates are changed and when conservation reduces demand and consequently revenue.

Pricing methods that encourage water conservation first require the use of meters to measure the amount of water used by individual customers and to charge for that water based on a rate schedule. When a customer is billed based on volume consumed, this is called *volumetric* billing. Where meters do not exist, customers are charged a flat rate, assessed at the same price for everyone in the customer class irrespective of how much water is actually consumed. When meters were eventually installed in many communities, water was commonly priced using a *declining block rate,* where customers are charged less per unit of water as consumption increases. This type of price structure discourages water conservation by rewarding high water users with a lower unit rate. Another common pricing mechanism has been the *uniform commodity rate,* where customers are charged the same unit price per gallon consumed regardless of the amount consumed.

The trend in many parts of the United States has now been toward *inverted, increasing,* or *inclining block rates.* These rates are also called *tiered block rates.* When the unit price for water increases as consumption increases, a signal is sent to customers to conserve, and most customers do respond to that price signal. The degree to which customers react to a change in price is called *price elasticity of demand.*

Residential water usage is largely a function of basic demographics, particularly household size, property size, and income. Nonresidential water use and outdoor residential water use (such as summer lawn watering) are generally more price responsive than indoor water use (Beecher 2009).

The first tier of a utility's water rates covers minimal water usage for a typical household at the minimum reasonable price; the subsequent tiers should be priced significantly higher (greater than 50 percent) than the prior tier to provide a recognizable price signal. Usually three to four tiers are adequate for an effective residential rate design (Whitcomb 2005, Alliance for Water Efficiency 2009).

In general, a five percent decrease in water consumption following a rate increase is a reasonable assumption. Long-term savings are usually lower than initial savings, because the effect of the price increase eventually wears off. Rates should be adjusted at least every few years to keep pace with consumption patterns as well as changing utility revenue requirements.

Significantly higher rates during peak use periods for customer use above a baseline is known as a *seasonal rate* and can be helpful in making customers aware of their contribution to peak demand. Seasonal changes in water rates can discourage water consumption during peak use periods.

All utilities, in coordination with the appropriate regulatory agency, should develop and adopt a *drought rate structure* appropriate for their service area. This would be implemented immediately upon declaration of a water shortage by the appropriate regulatory agency. Instead of a separate rate structure implemented during water shortage, the drought rate could be a surcharge added to the utility's existing rate structure. Drought rates can be tiered so that they send strong price signals.

Keeping base rates low and possibly designating rates for low water use customers as *lifeline rates* will generally remove most objections to rate increases. A more advanced type of rate structure, used successfully by several larger utilities, is called *budget-based water rates*. Information is now available on these rates from the Alliance for Water Efficiency (AWE 2008) and from a report recently released by the Water Research Foundation, *Water Budgets and Rate Structures: Innovative Management Tools.*

The major cost to modify water rates is the staff time required for the utility planner to develop the new rates and to predict their revenue impacts.

BILLING SYSTEMS

Billing data can provide before and after information on demand reduction attributable to specific water conservation measures (Figures 3-4 and 3-5). The utility manager should instruct the finance department to save billing data for a period of at least five years to allow for future analysis. Electronic data storage should make it easier to store.

To allow analysis of water savings before and after water conservation measures, the billing system should ideally be able to produce data in an easily analyzable form, for instance as Excel spreadsheets.

The water conservation manager should seek to develop relationships with the finance and the customer service departments, and request to be included on the committee to select a new billing software package, when that occurs.

At the next opportunity to select billing software, the ability to identify different customer classes through billing information should be included, for example a field for NAICS codes.

San Antonio Water System

P.O. Box 2990
San Antonio, Texas 78299-2990
(210) 704-SAWS (7297)

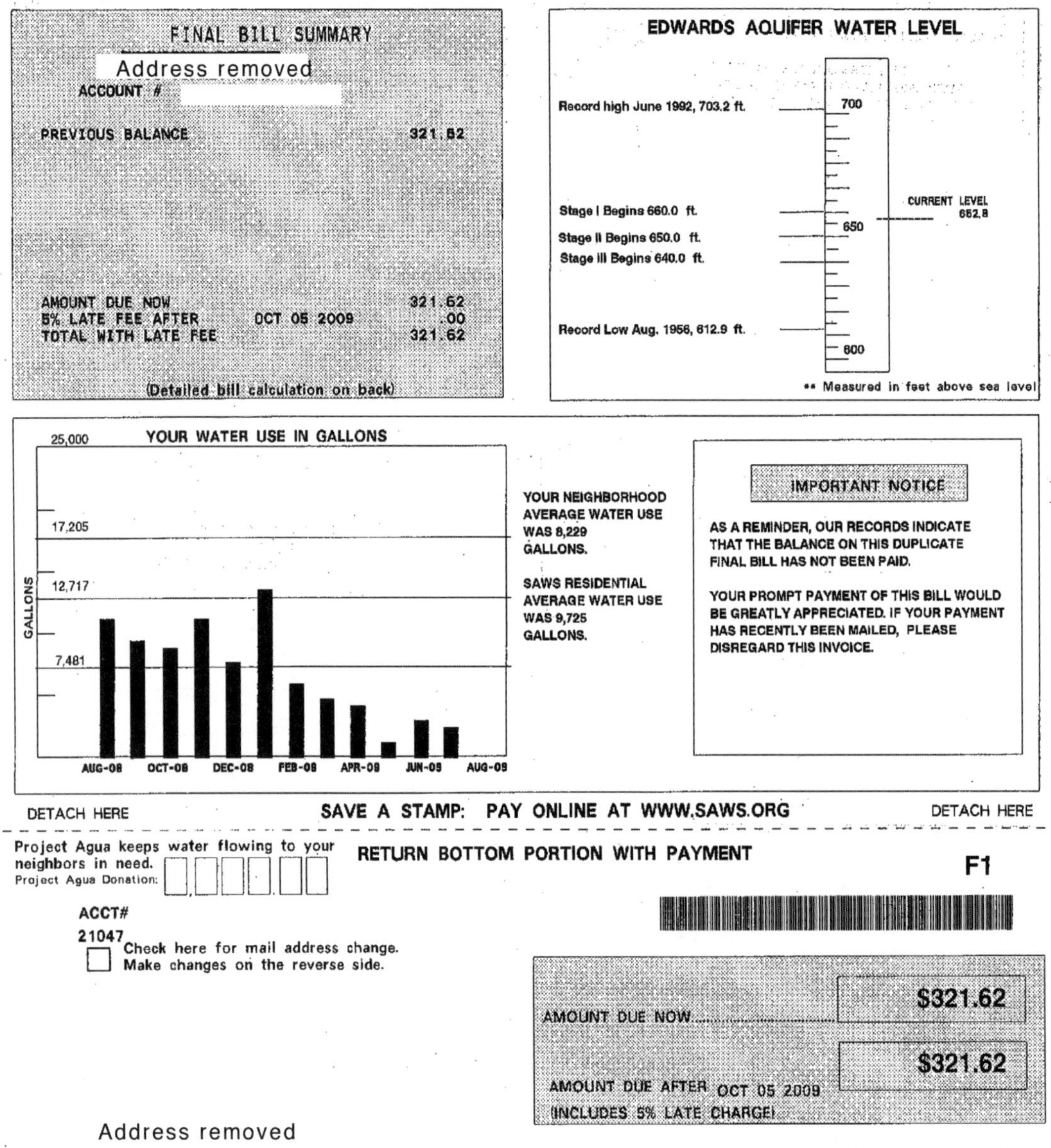

FINAL BILL SUMMARY

Address removed

ACCOUNT #

PREVIOUS BALANCE		321.62
AMOUNT DUE NOW		321.62
5% LATE FEE AFTER	OCT 05 2009	.00
TOTAL WITH LATE FEE		321.62

(Detailed bill calculation on back)

YOUR NEIGHBORHOOD AVERAGE WATER USE WAS 8,229 GALLONS.

SAWS RESIDENTIAL AVERAGE WATER USE WAS 9,725 GALLONS.

IMPORTANT NOTICE

AS A REMINDER, OUR RECORDS INDICATE THAT THE BALANCE ON THIS DUPLICATE FINAL BILL HAS NOT BEEN PAID.

YOUR PROMPT PAYMENT OF THIS BILL WOULD BE GREATLY APPRECIATED. IF YOUR PAYMENT HAS RECENTLY BEEN MAILED, PLEASE DISREGARD THIS INVOICE.

DETACH HERE SAVE A STAMP: PAY ONLINE AT WWW.SAWS.ORG DETACH HERE

Project Agua keeps water flowing to your neighbors in need.
Project Agua Donation:

RETURN BOTTOM PORTION WITH PAYMENT

F1

ACCT#
21047

Check here for mail address change.
Make changes on the reverse side.

AMOUNT DUE NOW	$321.62
AMOUNT DUE AFTER OCT 05 2009 (INCLUDES 5% LATE CHARGE)	$321.62

Address removed

SAN ANTONIO TX 78233-4537

334935261449900032162000321620

If you have any question about your bill, please call (210) 704-SAWS or write to: SAWS Customer Service, P.O. Box 2449, San Antonio Texas 78298-2449. SAWS appreciates the opportunity to serve you.

Figure 3-4. Water bill from San Antonio Water System is a good example of providing information on the bill to help the customer make conservation choices. (Graphic: San Antonio Water System)

CURRENT BILL CALCULATION

LAST SERVICE DATE: AUG 19 2009
RESIDENTIAL/ICL SERVICE ACCOUNT: 13 4935 261449 9

PREVIOUS BALANCE 321.62

TOTAL CURRENT CHARGES 321.62

HOW TO READ YOUR METER

0 0 3 2 1 X X

YOUR METER HAS A NUMBER DISPLAY SIMILAR TO THE ONE PICTURED ABOVE. (DISREGARD NUMBERS WITH A BLACK BACKGROUND.) SUBTRACT THE PREVIOUS METER READING (PROVIDED ON YOUR STATEMENT) FROM THE CURRENT ONE. THE RESULT WILL BE THE AMOUNT OF WATER USED SINCE THE LAST READING (IN HUNDRED CUBIC FEET). TO CONVERT TO GALLONS, MULTIPLY THIS AMOUNT BY 748.1. NOTE: WASTEWATER-ONLY AND STORMWATER-ONLY CUSTOMERS WILL NOT HAVE A METER READING SHOWN.

SUBTRACT PREVIOUS READING FROM CURRENT: 2-1=1

CONVERT TOTAL TO GALLONS: 1X748.1=748

Know when to turn it off.
saws.org

Last Digit of Street Address	Watering Day
0 or 1	Monday
2 or 3	Tuesday
4 or 5	Wednesday
6 or 7	Thursday
8 or 9	Friday
No watering on weekends	

Watering Hours:

Stage 1: Midnight - 10 a.m. and 8 p.m. - Midnight
Stage 2: 3-8 a.m. and 8-10 p.m.
Stage 3: Same hours as Stage 2, every other week
Stage 4: Drought surcharges may aply for excess water use

BexarMet water customers who receive sewer services from SAWS are now billed directly by SAWS for this service. Direct billing allows us to be more responsive to our wastewater-only customers. SAWS also bills for stormwater on behalf of the City of San Antonio. For questions or concerns about wastewater or stormwater billing call SAWS at 704-SAWS

Payments mailed as of Nov. 1, 2006 will be processed electronically.

San Antonio Water System

Para recibir su estado de cuenta en español favor de llamar 704-SAWS

CHOOSE A BILL PAYMENT THAT WORKS FOR YOU!

Pay by Phone: 704-SAWS(7297)
Pay Online: www.SAWS.org
Pay in Person:

CUSTOMER SERVICE LOCATIONS

Monday - Friday

---803 Castroville Road(Las Palmas) 8:00 am - 5:00 pm
---915 W.W. White Rd(Lone Oak) 8:00 am - 5:00 pm
---2800 U.S. Hwy 281 North 8:00 am - 5:00 pm

MAILING ADDRESS CHANGE

Number Street Apt#

City State Zip

New Home Phone: ( )

Figure 3-5. Water bill from San Antonio Water System is a good example of providing information on the bill to help the customer make conservation choices. (Graphic: San Antonio Water System)

Requesting regular demand reports identifying high water use customers will allow the water conservation manager to target high water users and follow the progress of programs.

As well as a source of data for future analysis of cost-effectiveness of water conservation programs, the water bills are an important means of communication with customers. The shorter the billing cycle, the more frequent the reminder to customers of the cost of water. Monthly billing is considered more effective than bimonthly and quarterly because it provides immediate information to the consumer that water usage may be high. If a bill is not received until the irrigation season is over, there is no opportunity for the consumer to adjust their consumption habits and therefore save water.

Information on the unit cost of water can assist customers in reducing use by making cost effective conservation choices. Customers also may have no idea that they use more water than the average customer, and thus the comparison between the customer and the average can be included on bills. Appendix C contains suggestions on what to include in "Informative Billing."

FUNDING THE WATER CONSERVATION PROGRAM

In many parts of the US, state funding is heavily biased toward building new facilities rather than in promoting conservation. Water conservation groups continually lobby for cost-effective water conservation measures to be eligible for cost-share funding proportionately with other alternative systems. They point out that conservation creates "new" supply, while reducing the need for new infrastructure. Again, the term *capacity buy back* is appropriate. Conservation yields water, which, when cost-effective, costs even the small to medium-sized utility far less than new supply options would be.

However, finding the funding for a conservation program is often difficult at a small utility. At the state level, conservation projects should be eligible for USEPA's Drinking Water and Clean Water State Revolving Funds, although each state currently differs in this eligibility. Until the American Recovery and Reinvestment Act of 2009, green infrastructure projects did not receive much official recognition in clean water and drinking water programs. This act mandated a 20 percent set aside for green projects such as energy efficiency, water efficiency, or innovative environmental projects Continued federal support for such projects appears likely.

Justifying the increased funding for water conservation is best supported by documentation of cost-effective savings showing that the water yielded is a bargain. In the earliest phase, a water conservation program will probably focus on education and may not have any programs that show a quantitative water savings. As budgets increase, or as the water shortage grows, retrofit programs that allow quantitative demand reduction can be carried out, which will document the benefit of the program. A good analysis of results of an outdoor program is sometimes difficult, due to unpredictability of the weather and many other variables. The small and medium-sized utility can rely on cost-effectiveness studies done by large utilities, some of which are referenced in chapter 4 (California Department of Water Resources 2008, California Urban Water Conservation Council 2005, Mayer et al. 2000).

Table 3-1 shows water conservation measures suitable for small and medium-sized utilities. Measures on the left are quantifiable in terms of water saved through metering and billing records. Those on the right are generally considered too difficult to quantify because of their breadth and the variety of factors that could also be affecting usage at the same time. For estimate of water saved by the measures on the left, see Table 4-1 (Modified from table by Kathy Scott, Southwest Florida Water Mgt. District).

Most conservation programs are funded through the utility's annual operating budget. Some utilities in fast growing areas have switched funding of conservation programs to the utility's capital budget, which is predominantly derived from bond funding or new connection fees. Bond funding is a preferred option, as it spreads the costs of the conservation program across the years that the savings will actually occur, and thus will not result in a cash drain for the utility during the first year of the conservation program. Connection revenues (also referred to as impact fees, tap fees, or water connect fees) typically pay for development of new resources and infrastructure, and thus can be used to fund conservation programs as well. In this way, newcomers are paying for the programs that help make water available for them, and current residents like this approach because the costs for new "supply" are borne by the new residents. However, connection revenues can dry up in slow growth periods and should not be considered as the sole funding source for conservation.

If a utility has a tiered rate structure and many high water use irrigation customers, the utility can also direct the funds collected from the upper tiers of the rate structure into the water conservation program budget. Many utilities have done this with great success. The advantage is that those customers who put stress on the supply system through their high seasonal water use are funding the necessary costs for the increases in water supply for the utility. All or a portion of this revenue may fund retrofits for older homes or whatever water conservation project is found to be most cost-effective (Chesnutt 1997, Johns 2007).

Some utilities charge a water rate surcharge on the customer bill, with 50-100 percent of revenue generated from this surcharge going back to the customers in the form of rebates and incentives. Reduction in wastewater treatment costs might be planned and the savings transferred to funding for the water conservation effort. Another funding source utilized by some utilities has been fines for pollution or watering restriction violations. In some areas, energy and water utility partnerships have been formed to co-fund programs. Incentives can be based partly on the water savings and partly on the energy savings. Distribution of low-flow showerheads has been done in this way. The California Urban Water Conservation Council's pre-rinse spray valve installation program for restaurants and Seattle Public Utility's WashWise clothes washer rebate program have both been co-funded by energy utilities.

WATER CONSERVATION COORDINATOR TRAINING

As a utility's water conservation program grows, the point person for the program may need to receive more training. Water conservation workshops are presented by AWWA twice a year at annual conferences. The information presented at these workshops and the networking opportunities will be invaluable. The utility's state section may also have a water conservation group and may offer training workshops. The AWWA WaterWiser website, www.waterwiser.org, and Alliance for Water Efficiency website, www.allianceforwaterefficiency.org, both publicize information on upcoming

Table 3-1. Comparison of Quantifiable and Not Easily Quantified Measures

Quantifiable	*Not Easily Quantifiable*
Residential Indoor	**Public Information/Giveaways**
Faucet Aerators (installation during audits)	Printed Materials
Leak Detection and Repair (by staff)	News Media
High Efficiency Toilets (rebates)*	Faucet Aerators (giveaway, installation not confirmed)
High Efficiency Clothes Washer (rebates)*	Leak Detection and Repair (print and web information only)
	High Efficiency Toilets (information)
Residential Outdoor	High Efficiency Clothes Washer (print and web information)
Xeriscape (turf replacement)*	Xeriscape (workshops, print literature, and demonstration gardens)
Landscape Irrigation Audit, w/ timer resetting, leak detection & repair	
Rain Sensor (giveaway or rebate—summer rainfall regions only)	**Regulatory/Economic**
Rain barrels (giveaway)	Water Conserving Rate Structure
	Watering Restrictions Ordinance
Commercial, Institutional, and Industrial	Coordination of enforcement
Restaurant Pre-rinse spray valve (giveaway w/installation)	Reuse Ordinance
Industrial Water Audit w/ or w/o rebates*	Plumbing Ordinance
Repair and Maintenance*	Landscape Ordinance*
High Efficiency Toilet and Urinal Rebates*	
Commercial Clothes Washer Rebates*	
ET (SMART) Controller or Soil Moisture Sensor Rebate*	

Note: * means appropriate for larger utilities or maximum programs of Checklist 1 in Chapter 5.

conferences and workshops. Water Conservation Certification programs, with accompanying training, are in place or under development by AWWA's Cal-Nevada Section, Pacific Northwest Section, and others.

The first Water Conservation Technician A.S. Degree Program in the country is at Lane Community College in Eugene, Oregon, www.nweei.org/water-conservation-tech.html. The utility may wish to contact this program to seek a qualified graduate to serve as a water conservation coordinator or take advantage of distance learning opportunities that may be available. In addition, local colleges and universities are all beginning to offer courses in water conservation and water resource management. Local college course listings should be checked.

Details about the role of the coordinator and this person's work with other members of the utility and the public appear in Chapter 7.

PUBLIC INFORMATION

Public information can be used to motivate voluntary customer conservation. It can be used to achieve both peak-use and year-round reductions and is particularly useful in drought years. Residential, commercial, and industrial customers can be targeted individually to accomplish conservation objectives specific to those sectors.

A public information program should raise awareness of water supply resources, including supply availability, treatment, and distribution issues. At the same time, it should educate customers about wasteful water use practices, such as over-watering lawns, not running full loads in dishwashers and clothes washers, and allowing leaks. The utility should set good examples for the public in all utility buildings and municipal landscapes, and this can be publicized.

The utility must be able to commit a certain amount of staff time and resources to a public information program for it to be effective. The size of the program is limited only by the utility's budget. Some larger utilities with aggressive programs budget $1 per person per year for this effort alone.

Cost sharing with the wastewater utility or other local water agencies in the same television and newspaper media area can reduce costs. Regional programs, because of economies of scale, can be very effective in delivering a conservation message to customers at far lower costs to the individual utilities than would be possible if they did the programs alone.

Advantages of collaboration include

- Avoiding duplication of effort
- Providing regional consistency
- Reducing costs, particularly for public education
- Achieving greater public visibility for programs

Examples are the Regional Water Providers Consortium in the Portland, Oregon area (www.conserveh2o.org), the California Friendly program of the Metropolitan Water District (www.bewaterwise.com), the Saving Water Partnership in the Seattle area (www.savingwater.org), and the Water Conservation Alliance of Southern Arizona (Water CASA, www.watercasa.org).

Components of a successful public information program typically include (Maddaus, 1987a)

A statement of program purpose. The public must be informed about the need for conservation. The utility's water supply situation should be explained clearly, and the specific objectives for the program should be stated (for example, peak-use reduction during summer months to avoid having to build a costly storage reservoir). This information can be included in the annual Consumer Confidence Report or other printed report and made available on the utility website.

A theme for the program. This may be a slogan or logo that will be used consistently that will become readily identifiable by the public. A recent Denver Water campaign was "Use only what you need." Other slogans are "Save Water, Save A Buck" of the Metropolitan Water District of Southern California, "Water Matters" of the Southwest Florida Water Management District, "Water Use It Wisely" in the State of Arizona, and "Slow the Flow" of the Center for Resource Conservation in Colorado.

Development of a campaign strategy. Public education is most effective as an ongoing aspect of the utility's overall utility conservation program. The public will soon forget messages that are only delivered once. The first year of the public education program should be thought out completely and budgeted adequately. The first year may justify more effort and resources than subsequent years, when a maintenance-level approach can be taken.

Identification of target groups. The utility should identify specific sectors of the audience and aim efforts specifically toward these groups. Potential target groups

The American Water Works Association has bill inserts available for quantity purchase. These include: *Slow the Flow,* outdoor conservation, *How Low Can You Flow,* indoor conservation, *Water Conservation at Home, Be a Leak-Seeker, 25 Things You Can Do to Prevent Water Waste, Five Basic Ways to Conserve Water,* and *Conserve Water* stickers. Visit www.awwa.org or call 800-926-7337.

include single-family homeowners – specifically high water-using single-family customers, apartment dwellers, business and industry (particularly large users), and local governmental agencies. A special focus might be on irrigation customers.

Formation of a water conservation committee. A committee can include elected officials, local business people, interested citizens, agency representatives, and concerned local groups. The function of the committee is to provide feedback to the utility about the public information aspects of its conservation plan, to develop new materials and ideas about public information, and to support conservation in the community. An example of a committee needing specific stakeholder input is the committee to develop a landscape ordinance – landscape professionals and potentially builders and developers should be involved.

Identification of techniques and resources. Communication can be accomplished through print media, radio, television, websites and blogs, and a speakers' bureau. Print media includes utility bill stuffers and paid newspaper advertisements. Press releases may yield free newspaper articles. City newsletters also provide free public information opportunities. Home Owner Association newsletters may also print articles based on press releases. Free radio and television public service announcements may be available. The utility staff person should choose the methods most likely to reach particular target groups. Volunteers and youth organizations may be available to help distribute educational materials. AWWA has public educational materials, such as bill inserts, available for purchase. Larger utilities may share their materials, and regional coalitions can reduce the cost of developing materials.

Updates and feedback. The utility staff should make an effort to keep public information program timely and up-to-date, by including new materials and ideas every year. Obtaining community feedback about the program's effectiveness is essential.

As a conservation measure in its own right, public information is likely to save the least amount of water when compared to the other measures discussed in this guidebook. When public information is the only conservation measure offered by a utility, water savings range from 2 to 5 percent during noncrisis periods. More typically, public information generally plays a supporting role to other conservation measures and is designed and conducted to support rebate programs, free showerhead giveaway campaigns, and free landscape audits. Although its direct effects alone usually cannot be measured, public information is needed to support consumer behavioral/lifestyle changes that may need to occur (such as accepting low water-use landscaping).

The public information campaign should be monitored on a yearly basis to assess progress toward the utility's goals. The goal of permanent behavioral modification is difficult to document. However, a quantifiable goal would be to closely monitor the increase in public support and media attention. The latter will assist in gaining the attention of elected officials, which increases political support of ordinances and increased funding.

Figure 3-6. Homepage of the Portland area Regional Water Providers Consortium as of December 2009. Site includes how-to videos on fixing toilet leaks among other useful information. (Courtesy: Regional Water Providers Consortium)

The book *Fostering Sustainable Behavior* is an excellent reference, with web resources and a forum where educators share experiences (McKenzie-Mohr and Smith 1999.

Appendix C contains examples of public education materials.

IN-SCHOOL EDUCATION

Educating youth about efficient water-use practices both establishes habits for the future and allows children to bring messages home to their parents. If materials used are low-cost (for example, informational pieces created by the utility about the utility, resources borrowed from other water providers, and/or internet research), these programs can be inexpensive to implement.

School career days are good opportunities for school visits by utility staff and offer an opportunity to develop relationships that can lead to more involved programs. Water-efficient demonstration landscapes developed by schools may be successful in securing interest of parents in adopting some of the same practices at home. The sustained interest of at least one key teacher and the principal, plus efforts by the utility, is essential for a school demonstration landscape to succeed.

Water conservation poster contests have been used successfully by many utilities, working with art teachers, and provide an opportunity for school visits by utility staff. Ways to honor the winners and receive good community interest are putting the winning images on T-shirts, billboards, water bottles, or calendars. However, those methods are costly. Winning images can be most economically displayed on the utility website.

Conservation and water-use education can be incorporated into the science curriculum for some grades. Some water management districts have developed region-specific in-school materials. Others have funded existing educational programs, for example the Project WET teacher training and curriculum program (www.projectwet.org). Some state agencies will assist utilities to put on workshops for teachers through Project WET.

The USEPA and USGS both have water conservation educational activities. AWWA's WaterWiser website has compiled links to the latest materials.

The utility's responsibilities include

Obtain permission from school authorities to introduce education program materials.

Identify a teacher or teachers who are willing to work with the utility to ensure conservation information can be provided in the classroom.

Provide a teacher guidebook or teacher education materials on the subject and coordinate teacher training, including offering continuing education credits/units (CEUs).

Estimate the number of schools, teachers, and students participating in the program; prepare and distribute the necessary materials.

Offer tours of water facilities, participate in school information fairs, and do follow-up on teacher response to the materials provided.

A well-designed in-school education program offers the longest-range conservation savings in that it creates a foundation for a change in perception in the community.

Review Consumer Conservation Measures

SUMMARY:

Describes potential conservation measures for reducing consumer water usage:

- *Residential, indoor*
- *Residential, outdoor*
- *Commercial*
- *Institutional (public facilities)*
- *Industrial*

In this chapter, the focus is on technologies that improve water efficiency. Measures that are widely used by utilities with supply problems are profiled, and an approach to each measure feasible for small and medium-sized utilities is suggested. Large utilities can provide financial incentives to customers through rebates to accelerate the market penetration and consequent demand reduction from these measures. Small- and medium-sized utilities can often also provide cost-effective incentives to customers and at least provide information to customers on the savings they will achieve.

Table 4-1 summarizes the potential water savings for some of the measures utilized by larger utilities. Figure 6-1 suggests a general approach to selecting measures based on an individual utility's customer base, which was profiled on Worksheets 2A-2D.

Although outdoor water use is often the largest residential use of water and is the most discretionary, most utilities focus first on indoor water use because it is simpler and yields quick results. For indoor use, reductions depend less on customer behaviors and come almost exclusively from the retrofitted fixtures. For outdoor water use, customer expectations of landscape appearance and customer behavior in operation of irrigation systems are highly variable and difficult to influence toward greatest efficiency. Irrigation technologies are also more complex than technologies for indoor fixtures, more prone to errors of design, and more likely to go into misalignment and disrepair.

As updates to information presented in this chapter, a utility can find current information from the California Urban Water Conservation Council Best Management Practices (Koeller 2008), Alliance for Water Efficiency Resource Library at www.allianceforwaterefficiency.org, and USEPA WaterSense program, www.epa.gov/watersense.

RESIDENTIAL MEASURES, INDOOR

In 1999 the Awwa Research Foundation (now known as the Water Research Foundation) published the *Residential End Uses of Water Study*. This study of more than 1,100 households found that conserving homes with low flow fixtures used 30 percent less water indoors than nonconserving homes (Mayer et al. 1999). When this data is compared to indoor water use data from a 1984 study (US Department of Housing and Urban Development 1984), a 10 percent reduction in indoor water over that 15-year period can be seen (Vickers 2001). These decreases in indoor use are due to improvements in efficiency of fixtures, as required by EPAct 1992 federal law. As another example of how water use has decreased, the amount of water currently used to meet the sanitation needs of 34 million people in the state of California is less than that used in 1980 to meet the needs of only 24 million people (Gleick et. al 2003).

Culminating the years of regulation and product testing effort, the USEPA WaterSense program, www.epa.gov/watersense, has been developed. It is a third party certification and labeling program for water using products, equivalent to the USEPA Energy Star program.

Indoor Water Audits and Leak Repair

The best strategy to identify customer conservation potential is an indoor audit, usually conducted by a utility staff member. Depending on staff capabilities, indoor audits may be costly in terms of staff time, and some utilities are deterred by liability concerns (although this can be easily handled with a customer liability waiver form).

As an alternative, a small or medium-sized utility can provide a flyer with fill-in calculation sheets to guide the customer in a self-directed home water audit. Additionally, a smaller utility may be able to provide a link to an online survey. Utilities can provide a link to the California Urban Water Conservation Council's H2Ouse site (www.h2Ouse.org), developed by the California Urban Water Conservation Council. As shown in Figure 4-1, this site is an excellent source of information for consumers on water use throughout a typical home, and also has a Water Budget Calculator. WaterSense also has an online calculator that can help the customer calculate payback time for retrofits at www.epa.gov/watersense.

However, many customers may not have easy Internet access and may appreciate the assistance of an auditor in their own home. Procedures for indoor, as well as outdoor, water audits appear in Appendix B.

The *Residential End Uses of Water Study* found that 13.7 percent of indoor use is from leaks, almost all related to toilets (Mayer et. al. 1999). Indoor use in a nonconserving

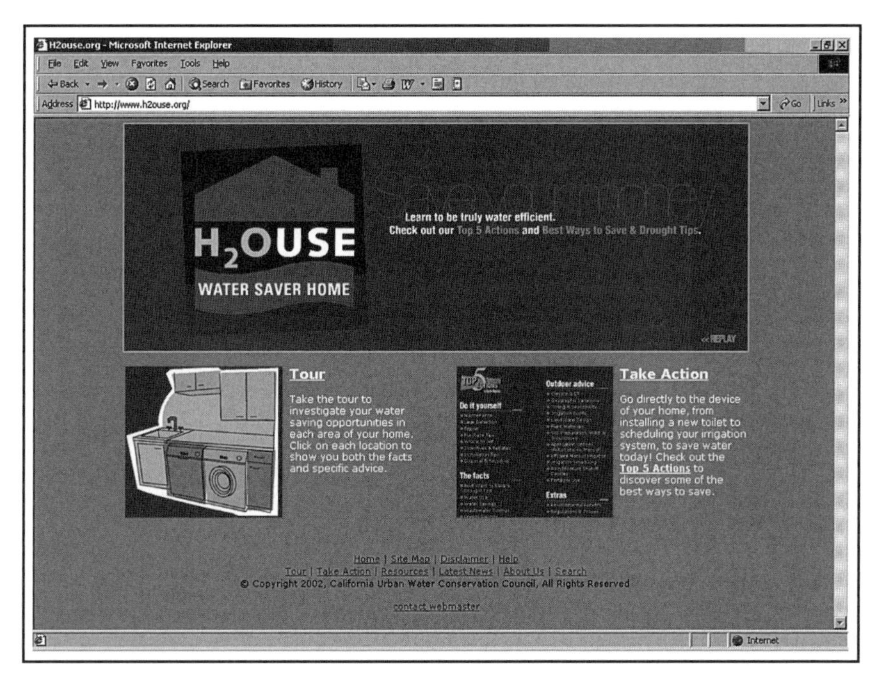

Figure 4-1. H2Ouse website, www.h2Ouse.org, to which any utility can provide a link, is a source for household water efficiency information. (Photo: Maddaus Water Management)

home averages 69.3 gallons per capita day, and savings from fixing leaks would be as much as 9.5 gallons per capita day. These leaks often consume more water than the amount used for flushing overall.

Many water customers allow leaks to continue because they do not have the skill to fix them and cannot afford a plumber. The utility can provide a staff member to repair leaks or develop a list of preferred contractors. For example, plumbers that have taken the GreenPlumbers courses (profiled later in this chapter) might be recommended. Having this service available is particularly valuable when customers call complaining about high water bills.

Elderly residents may not hear running toilets and other audible leaks, and many leaks are silent. Leaking toilets can be checked by using dye tablets and by observation of the refill tube. The tablets are dropped into the toilet tank; if a leak exists, the dye color will appear in the toilet bowl within about 20 minutes. A toilet that continually refills indicates that the float should be changed.

Utilities can use giveaway programs to distribute dye tablets, mailing or delivering them to customer homes, and providing information on their use. Drops of food coloring in the toilet tank can be similarly promoted. Some utilities tell their customers to test for toilet leaks when they change their clocks, twice a year. Regulatory programs can also require leaking toilets to be repaired or replaced at the time of home resale.

The USEPA WaterSense program has an annual program to publicize customer leak detection—"Take a Pledge to Fix a Leak! Week." Sample text for promotional material is available from the WaterSense website.

Residential Fixture Retrofit Kits

Retrofit kits can contain any combination of devices to reduce the water consumption of toilets, showers, and sinks. A typical kit (Figure 4-2), may contain two low-flow showerheads complete with Teflon tape, two toilet tank displacement devices for pre-1994 toilets, two toilet tank leak-detecting dye tablets, two bathroom faucet aerators, and a kitchen faucet aerator. The kits should contain installation directions for all devices along with literature on efficient water-use practices for both indoor and outdoor residential water use.

A small or medium-sized utility should seek opportunities to secure installation of the items in a retrofit kit, possibly by a utility staff member installing it as part of home water audit. In some communities, nonprofit organizations train youth to perform audits and minor retrofits. For multi-family housing, a staff member from the facility can be trained in installation by a utility staff member. Without a specific method to insure installation, retrofit kits must be considered basically a public information measure.

Retrofit kits, or low-flow showerheads and faucet aerators alone, are appropriate giveaways at workshops and open house events. Utilities can offer customers retrofit kits free upon request and advertise the kit availability on websites and in customer newsletters. In addition to allowing customers to order a full kit, a utility can create a menu of devices from which people can order. In multifamily retrofit programs, the kits are delivered to the building manager for the maintenance staff to install.

Toilets

Flushing the toilet is typically the largest indoor residential water use. Indoor water use was 69.3 gallons per capita day in a representative sample of 1188 homes in 12 cities across the U.S. and Canada (Mayer et al. 1999). These homes were randomly selected in 1996 and 1997, so some had older fixtures and others had 1.6 gallon per flush toilets specified by EPACT 1992. This information, from data logging studies conducted under the AWWARF funded *Residential End Uses of Water Study*, indicates that an average person at home flushes the toilet 5.1 times per day. Toilet flushing made up 26.7% of the total indoor water use in that study (Mayer et al. 1999).

Toilet use is an area where substantial water savings may be achieved. In studies of retrofitted homes in Seattle, WA, East Bay Municipal Utilities, CA and Tampa, FL, fixtures with efficiencies specified by current codes or better were installed replacing older fixtures. Indoor water use averaged 43.8 gallons per capita day (Aquacraft 2000, 2003a, 2003b). A comparison of the proportion of use by different fixture categories in older and newer homes appears in Figure 4-3. Toilet use earlier made up about a quarter of indoor water use, but was reduced to 18.5% of the indoor total following retrofit. Indoor usage is expected to decrease even more with retrofits of WaterSense labeled fixtures.

Because remodeling bathrooms is a popular aspect of home renovations and increases the value of a home, many customers retrofit on their own and can be provided information about efficient models and/or payback. Other customers are willing to replace a toilet if they have financial assistance.

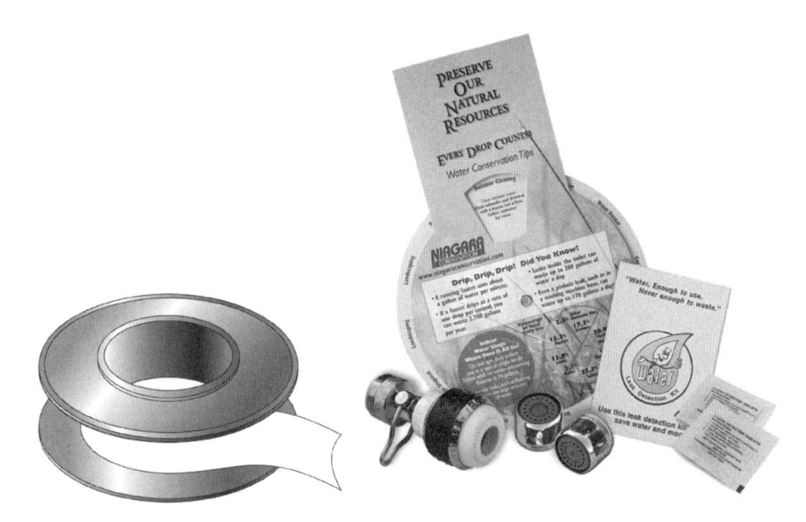

Figure 4-2. Water conservation retrofit kits, with optional energy efficiency items, are available packaged for distribution from several water conservation vendors. (Graphic: Niagara Conservation)

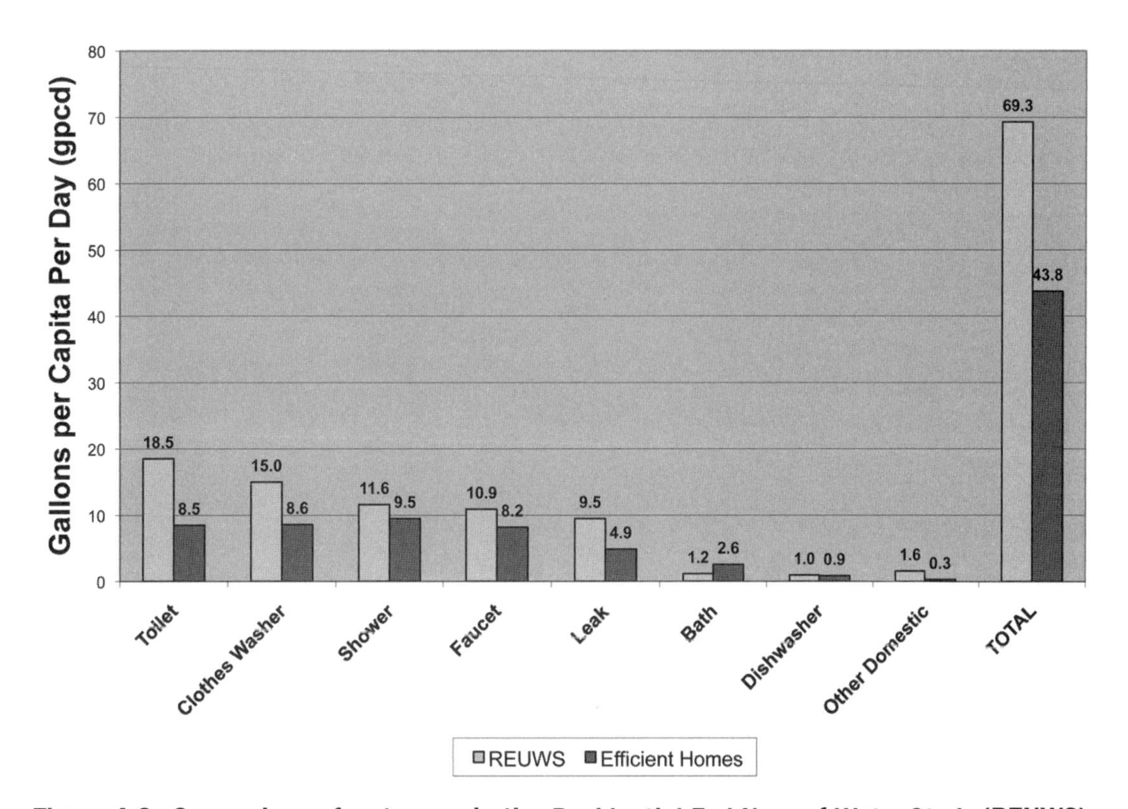

Figure 4-3. Comparison of water use in the *Residential End Uses of Water Study* (REUWS) and in homes in Seattle, Wash., Tampa, Fla. and the service area of East Bay Municipal Utilities, Calif., that were retrofitted to 1992 code or better (Aquacraft 1999, 2000, 2003a, 2003b)

High Efficiency Toilets (HETs).

Since the advent of the High Efficiency Toilets in 1999, it is in the utility's best interest to specify these models for any regulation or incentive programs, as they function exceptionally well and save 20 percent more water than the currently-required low-flush toilets using 1.6 gallons per flush (gpf). These high-efficiency fixtures typically use only 1.28 gallons per flush.

One of the primary methods of achieving these low flush volumes is dual-flush design, as shown in Figure 4-4. Pressing one button allows a flush of half volume for liquid waste, while the other allows a full flush. The average of the full volume (1.6 gpf) and half volume (0.8 gpf) works out to be approximately 1.28 gpf. In fact, even lower volumes are now being achieved, because the tank volume for the full flush in some models is 1.28 gpf (Koeller 2009). Dual-flush toilets are available in the U.S. from 30 different manufacturers.

In addition to dual-flush design, several other mechanisms have been developed to achieve the 1.28 gpf standard while providing satisfactory performance. HETs that have been tested by USEPA's voluntary water conservation labeling program, WaterSense, perform well and oftentimes better than toilets meeting the current code of 1.6 gpf (Koeller 2008).

Along with the third party testing of toilets for the USEPA's WaterSense program, another toilet testing report is available for information, the Maximum Performance (MaP) testing reports. Listings of toilets that have passed the MaP test are posted and ranked by score on the California Urban Water Conservation Council and Alliance for Water Efficiency websites. However, since the USEPA developed the WaterSense program in coordination with the California Urban Water Conservation Council and Alliance for Water Efficiency, the customer-friendly WaterSense site is now sufficient for reference by most utilities.

Figure 4-4. The dual-flush toilet, originated by Caroma in Australia, is a High Efficiency Toilet (HET) and now widely available from many manufacturers. (Graphic: Water Media Services)

Figure 4-5. Display at home supply store promoting WaterSense labeled HET toilets (Photo: USEPA Water Sense and Lowe's)

Retrofitting older toilets, 3.5 gpf or higher, with HETs can be promoted by regulation, financial incentive, or public education that encourages voluntary purchase. Many larger utilities provide $50–$200 rebates to homeowners who voluntarily retrofit using HETs, instead of retrofitting with the currently legally mandated 1.6 gpf low flush toilets. The customer can provide proof of purchase of a toilet from an accepted list (WaterSense certified) and be reimbursed for the rebate amount. These rebates can also be in the form of a voucher redeemable by a participating vendor. Some of the large home supply stores are participating with utilities in these programs.

A key element of these programs is prominent labeling of the WaterSense labeled toilets, as in Figure 4-5. When rebate funds are discontinued, customers are still likely to choose these more efficient toilets once the WaterSense label becomes widely known by the public, as is the USEPA's earlier Energy Star label.

The retrofit from a pre-1994 toilet using 3.5 gpf or more to an HET reduces per capita water usage by about 12 gallons per capita day. With only one person using the toilet, it reduces a household's water bill about 359 gallons. With three or more people using the toilet, over 1,000 gallons a month are saved, enough to translate into direct saving on a household's water bill. One word of caution: the toilet being retrofitted should be 3.5 gpf or higher. Replacing a 1.6 gpf toilet with an 1.28 gpf HET will likely not yield cost-effective savings. Information on age of homes in a utility service area, that relates to which volume of fixture was originally installed, has been compiled in Worksheet 2B.

Low-flow Toilets. This term refers to the generation of toilets that are specified by EPACT 1992 to use only 1.6 gpf. Replacement of the earlier generation 3.5 gpf toilets and the even earlier 5-7 gallon per flush toilets by these 1.6 gpf toilets saved 16–23 gallons per capita day (Chesnutt et al. 1992) and was an important conservation measure. However, these fixtures should no longer be included in utility rebate or retrofit programs, because they have been required by federal and state building codes since January 1, 1994, and because high performing HETs are now widely available. In 2001, nearly 50 percent of 1.6 gpf toilets were found to have a flush volume greater than 1.6 gpf (Canadian Mortgage and Housing Corporation 2001). Furthermore, the MaP testing program in 2002 and 2003 found that only about half of tested models successfully performed with one flush (Gauley and Koeller 2003).

Flapper replacement. The flapper is the moving part of the toilet flush valve that seals water into the tank and allows water to exit the tank when it is flushed. Flappers deteriorate over time as a result of in-tank cleaning products and disinfection chemicals used by utilities. A warped or poorly fitting flapper may cost hundreds of dollars a year in leaks (Bracciano et al. 2004).

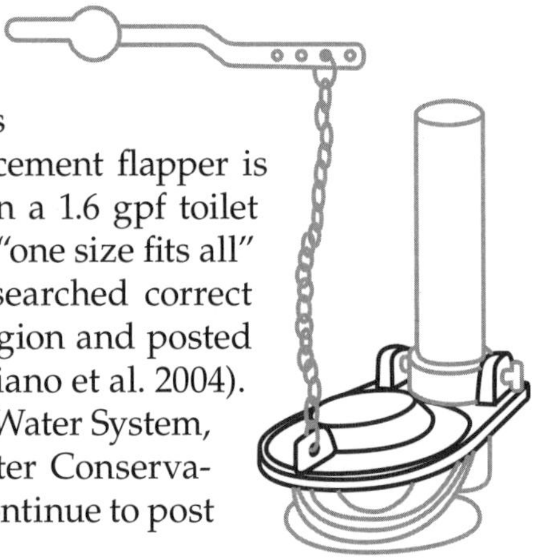

A dye test will reveal if a leaky flapper needs to be replaced. However, finding the correct replacement flapper is not always easy. The wrong replacement flapper on a 1.6 gpf toilet can convert it to a 3.5 gpf toilet. The perception that "one size fits all" can cause serious problems. Tampa Bay Water researched correct flapper models for the toilets widely sold in that region and posted these model numbers on a dedicated website (Bracciano et al. 2004). Several other utilities, for example the San Antonio Water System, provide similar information. California Urban Water Conservation Council and the Alliance for Water Efficiency continue to post new information on flappers.

Currently sold flappers use new elastomeric compounds developed in the 1990s. In 2004, the California Urban Water Conservation Council, with funding from many of its utility members, commissioned a field study of 900 homes to determine the extent to which these flappers were withstanding bowl-cleaning tablets. The findings indicate that the flapper failure (leakage) problems of the 1990s had largely disappeared with the newer flappers. By 2005, flapper durability and marking requirements had become incorporated into national product standards (Koeller 2009).

Retrofit items to reduce toilet flush volume. These devices are considerably less expensive than assisting customers with actual toilet replacement. Distribution of retrofit kits can potentially save water and increase awareness about water use and the importance of using water efficiently.

- **Toilet Tank Displacement Devices.** These devices reduce the amount of water used in each flush and save about 0.7 gpf (Konen et al. 1992). Previously, placing a brick in the toilet tank was used as a displacement device, but this is currently discouraged because bricks disintegrate. Utilities can encourage voluntary use of these devices or can institute device giveaway programs. Because residents can easily remove these devices, however, their long-term effectiveness is questionable. They also have a limited life, estimated at 3–5 years. A utility can also encourage customers to place their own sand-filled bottles into the toilet tank, and if they interfere with toilet functioning, later remove them. These devices should never be used with modern toilets using 1.6 gallon per flush or less.
- **Variable Flush Time Devices.** These devices operate by altering the force acting on the toilet tank flapper valve. This shortens the cycle time, which reduces the flush volume. Also called early closure devices, they commonly use a float attached to the flapper valve. They are relatively easy to install but require careful adjustment by the installer to maximize water savings while providing a satisfactory flush. Laboratory tests indicate a savings of 1 gpf (Konen et al. 1992). These should not be used on 1.6-gallon toilets.

Figure 4-6. With newer faucets, faucet aerators are not difficult to remove for cleaning or replacement. Typically two washers hold the faucet aerator in place. The aerator is inscribed with the flow rate. A maximum of 2.2 gpm has been the standard since 1994. USEPA WaterSense labeled aerators have flow rates of 1.5 gpm or less. (Photo and Graphics: Water Media Services)

Low-flow Faucets

Faucets account for more than 15 percent of indoor household water use—more than 1 trillion gallons of water across the United States each year. Low-flow faucets reduce water flow using a restrictor or aerator, as shown in Figure 4-6. With bathroom faucet usage 0.5 to 5 minutes person per day (Mayer et al. 1999), a faucet aerator can save 1.2-2.5 gpm, while reducing splashing and creating appearance of greater flow.

While older nonconserving faucets have flow rates up to 5 gpm, EPAct 1992 set the maximum at 2.2 gpm for all residential bathroom lavatory faucets. Bathroom faucets that use as little as 0.5 gpm have been well received by residential customers. Kitchen faucets that flow at 1.5 gallons to 2.0 gpm are also well received. Commercial faucets are specified by the model plumbing codes to use no more than 0.5 gpm.

Bathroom faucets and faucet aerators were among the first products to be labeled by the USEPA's WaterSense program, and the WaterSense website (www.epa.gov/WaterSense) has an updated list of models that have passed this third party testing. The WaterSense specification sets the maximum flow rate of faucets and aerators at 1.5 gallons per minute (gpm), tested at a flowing pressure of 60 pounds per square inch, which is the common water pressure in most households. A minimum flow rate of 0.8 gpm, tested at a flowing pressure of 20 psi, is also specified to ensure performance across a variety of different household conditions. The USEPA estimates that if every household in the United States installed a WaterSense labeled faucet or aerator, there would be a savings of more than 60 billion gallons of water annually.

Faucet aerators can be purchased by a utility economically in bulk to be given away to water customers individually or as part of retrofit kits. Note that removing the old aerator may be difficult for many customers. Directions for removal, including wrapping the faucet with a towel, and gently loosening the aerator with a wrench, should be included with kit distribution. Aerators can also be installed during indoor water audits, at customer request.

Low-flow Showerheads

Low-flow showerheads, as shown in Figure 4-7, reduce the amount of water used during showering by restricting the flow through the showerhead. Other benefits of using low-flow showerheads include lower energy costs from reduced water heating and reduced wastewater flows, both of which help consumers lower household bills. EPACT 1992 standards effective January 1, 1994 require showerheads to release no more than 2.5 gpm at 60 psi. Older showerheads have flow rates of 3–8 gpm.

The USEPA WaterSense program now has a high-efficiency showerhead standard for flow rates lower than that of EPAct, for both fixed and handheld showerheads. The 2009 WaterSense Draft Specification for Showerheads sets the maximum flow rate at 2.0 gallons per minute (gpm) at a flowing pressure of 80 pounds per square inch (psi). EPA worked with a variety of stakeholders to develop these criteria so that showerheads can be independently tested and certified for efficiency and performance. The utility should check the WaterSense website, www.epa.gov/watersense, for the final specification and for products that meet the strict WaterSense requirements for good performance. A major issue is danger of scalding in showers with two-handled non-compensating valves, commonly installed in older homes built prior to about 1987. Utilities that promote 1.5 to 2.0 gpm showerheads for retrofits assume a contingent liability with the possibility of users experiencing thermal shock or scalding accidents (Koeller 2009).

By reducing flow from 3 gpm to 2 gpm, water savings is about 1 gpm. Average showerhead use is 0.67 showers per person per day, and typical shower length is 8.5 minutes per shower, according to the *Residential End Uses of Water Study* (Mayer et al. 1999). Savings would be 5.7 gallons per capita day, or even greater if the old showerhead used more than 3 gpm. If the number of household residents is large or includes teenagers (or others who tend to take long showers), savings could be significantly greater.

Figure 4-7. Low-flow showerheads on display at the United States Green Building Council's conference Green Build 2008. Showerheads with flow rates of 2.0 gpm or less are appropriate for new construction. (Photo: Water Media Services)

A utility can purchase low-flow showerheads in large quantities from one of the several vendors specializing in water conservation device sales. A recommended method of showerhead giveaway that assures installation is called *showerhead exchange*. Through publicity, possibly tied to an open house or other water conservation event, customers are notified that they can receive a free low-flow showerhead if they bring in their old showerhead. Customers readily bring in their old showerheads, and the utility can with confidence log a specific quantity of water savings, knowing that the new showerhead will be installed.

In light of the fact that 2.5 gpm showerheads have been required since 1994, and a large proportion of showerheads may have been replaced since that time, the actual water savings value of a current showerhead giveaway, even with exchange, may be considered low (unless showerheads of 2.0 gpm or less are used, and then the scalding liability above is a factor). However, many customers welcome a new showerhead, if promoted properly, and the exchange can be tied with other water conservation publicity and outreach efforts to customers.

Informing customers on typical showerhead flow rates and the benefit of shorter showers can be part of public information, alone or accompanying the showerhead exchange. How a customer can calculate flow rate of his or her current showerhead can also be part of public information (See Residential Audit Procedure in Appendix B). Although any measured container can be used, plastic bags with printed measurements, as sold by vendors of water conservation devices, make flow rate measurement easy.

Water-efficient Clothes Washers

Water-efficient clothes washers use less than half the water of nonconserving clothes washers. An older clothes washer uses over 35 gallons per load. A family of four using a standard clothes washer will generate more than 300 loads per year, consuming 12,000 gallons of water annually. High-Efficiency Washers (HEW) can reduce water use by more than 6,000 gallons per year, saving energy, cleaning clothes better, and reducing fabric wear. Clothes washers are used at a rate of 0.96 loads per household day (Gleick et al 2003), and there are 0.82 machines per household (US Census Bureau 2005 in Cooley et al. 2007).

Replacing these appliances can also save energy because of reduced hot water use. In addition to using less water, ENERGY STAR® qualified washers extract more water during the spin cycle, which saves energy by reducing drying time.

Horizontal-axis (usually front-loading) washers, shown in Figure 4-8, are more efficient than conventional vertical-axis (top-loading) washers with agitators, in part because they need not fill the tub completely with water. New top-loading designs wet the clothes from above with sprayers and save energy and water over conventional top-loaders, but they may not clean clothes as effectively, according to Consumer Reports (Alliance for Water Efficiency 2009).

Because clothes washers have different drum capacities, ranging from 1.7 cubic feet to more than 4.2 cubic feet (with 3 cubic feet on average), it was necessary to develop a means to compare water use. Energy Star now includes a Water Factor (WF) in its ratings of products that use water, representing the quantity of water in gallons for

Figure 4-8. Front-loading clothes washer, Tier 3 of CEE, may use as little as 4.5 gallons per cycle. (Photo: Maddaus Water Management)

each cubic foot of drum capacity. The smaller the WF rating, the more water efficient the clothes washer (Alliance for Water Efficiency 2009).

The Consortium for Energy Efficiency (CEE), a nonprofit organization, has worked with water conservation professionals to initiate a Super-Efficient Home Appliances Initiative (SEHA). CEE maintains updated lists on clothes washer performance at www.cee1.org. CEE's Tier 3 has a water factor of 4.5 and below, Tier 2 has water factor of 6 and below, and Tier 1 has water factor of 7.5 and below.

Many water utilities that provide rebates for clothes washers focus only on those rated at the highest tiers by CEE. Seattle Public Utility's Saving Water Partnership has joined with other utilities in the Pacific Northwest in a WashWise Rebate Program. This program offers rebates of $50 for Tier 1, $75 for Tier 2, and $100 for Tier 3 clothes washers.

Assisting customers to replace their clothes washers will likely be beyond the financial capability of small and most medium-sized utilities unless the avoided cost of water is very high (Table 4-1). There is also concern that the clothes washer may be taken when a customer moves, unlike some of the other fixtures, and if moved outside the utility service area, the utility's water savings value of the rebate would be lost.

As an alternative, appropriate for small and medium-sized utilities, the utility can provide information for customers who are voluntarily purchasing a new clothes washer. The cost difference of buying an HEW over a conventionally sold machine will be paid back in 1 to 6 years (DeOreo et al. 2001). The utility can also promote the message to wash only full loads. In general, the washer drum should be full to the top. Research by Seattle Public Utilities confirms that the resident can save up to 2,000 gallons of water a year and save time by filling each load more completely (Broustis 2006).

Water-efficient Dishwashers

Retrofitting dishwashers has lower savings potential as a utility water conservation measure than retrofitting other indoor fixtures, mostly because residential dishwashers are a small part of residential water use and also because dishwashers have become so efficient. In 1999, dishwashers represented approximately 1.4 percent of typical residential indoor water use, and Americans averaged just 1 gallon per person per day using dishwashers (Mayer et al. 1999). Since 1999, automatic dishwasher use in the home has declined by 33 percent, and the percentage of total water use is now estimated to be less than 1.0 percent. Anecdotal information is that an increase in "eating out" has contributed to reduced use of dishwashers (Koeller 2008).

While older dishwashers average 14 gallons per load (1980-1990 data, cited in Vickers 2001), 6 gallons per load is typical of currently sold models. The average ENERGY STAR qualified dishwasher uses one-third less than other currently sold models, 4 gallons per load, and significantly less energy (www.energystar.gov).

Small and medium-sized utilities can provide information about benefits of ENERGY STAR dishwashers to customers in newsletters and direct them to the ENERGY STAR website.

Plumber Education

Creating alliances with plumbers in a utility service area can expand the effectiveness of water conservation efforts. The GreenPlumbers® program was first developed in Australia in 2000, as a result of a severe drought. "Engaging an army of plumbers" to help with water conservation efforts was the goal. The program started in the United States in September 2007, and now offers a series of high quality workshops throughout the country.

Prior to taking a GreenPlumbers course, most plumbers had never heard of the MaP testing of toilets or WaterSense. Having a "bad taste in their mouths" from the poor performance of the first generation of low flow toilets in the early and mid 1990s, many plumbers are distrustful of water conservation innovations. Through the GreenPlumbers course, plumbers learn which are the most water efficient and high performing fixtures to recommend. In addition to best toilet, faucet, and showerhead technology, plumbers in the workshops learn about reasons for water conservation including climate change, and learn about solar water heating and efficiency in hot water heating. A utility can sponsor plumbers to attend the workshops or simply make sure they are aware of the workshops. For further information, visit www.greenplumbersusa.com

Water Heating Issues

Water heating is the second largest energy use in new homes after space conditioning (heating and cooling), making up about 12 percent of a home's total energy requirement. There are four ways customers can cut their bill for water heating: Use less hot water (through efficient showerheads, faucets, clothes washers and dishwashers); turn down the thermostat on the water heater; insulate the water heater and pipes; or buy a new, more efficient water heater. The US Department of Energy is in the process of establishing an ENERGY STAR residential water heater program. A method to calculate energy utility bill savings from water conservation measures like low-flow showerheads, dishwashers and clothes washers is found in the AWWA M52 *Water Conservation—A Planning Manual* (Maddaus 2006).

The water wasted while waiting for water to get hot, particularly in showers, may be significant, which is an added benefit of low-flow showerheads. Tankless water heaters ensure that water does not run out; however they do not provide hot water to the user more quickly (Koeller 2007). Because tankless water heaters are marketed for "endless showers," they are not currently considered as contributing to water conservation savings.

As the average size of a new home has grown from the 1970s to present, water has had to travel farther from the water heater. New homes should be designed with "structured plumbing" to conduct hot water as efficiently as possible (Klein 2004). Since this approach is quite different from the conventional approach to plumbing layout, training of plumbers is necessary and has been incorporated into the Green-Plumbers courses.

Alternative systems to address water wastage in hot water supply include recirculating systems (on-demand and continuous), parallel pipe manifold systems, point of use heating, and wastewater heat recovery (Alliance for Water Efficiency 2009). The USEPA recognizes the Metlund OnDemand System as highly energy efficient (Chinery 2006).

OUTDOOR RESIDENTIAL MEASURES

Outdoor water use is vastly more variable than indoor use, ranging from 7 gallons per person per day to 52 gallons per person per day (Mayer et al. 1999). The majority of outdoor water use involves landscape irrigation. Approximately 59 percent of urban water use was found to be devoted to this discretionary use in the *Residential End Uses of Water Study* (Mayer et al. 1999), and in some regions preliminary data shows the percentage to be even higher currently (Mayer 2009). Outdoor water use ranges from 22–38 percent in humid climates such as Florida to 59-67 percent in the arid climates such as Arizona and California (Mayer 1999). The amount of water needed for landscape irrigation is determined by climate, plant material, soil type, and irrigation practices.

Overwatering is estimated to be as high as 30 percent, and just a 3-minute reduction of irrigation time can be the equivalent of several indoor measures (Dickinson and Gardener 2004).

LOW WATER-USE LANDSCAPING (XERISCAPE)

The name Xeriscape™ combines the Greek word root for dry (*xeri-*) with landscape. The term was developed and first promoted by Denver Water in the 1980s to encourage water conservation through creative landscaping (Dickinson and Gardener 2004). Terms with similar meaning are *water-wise landscaping* or *water-efficient landscape.* Throughout the United States, the state University Extension services are repositories of information on this type of landscaping and may have their own terms.

Xeriscape uses native or low water-use plants, emphasizes soil improvement and the use of mulches for soil water retention, reduces irrigated turf area, employs efficient irrigation system design, and applies overall landscape design to integrate all these elements. Lastly, it emphasizes maintaining the landscape and irrigation system appropriately.

Soil Preparation and Mulching

Good soil preparation and care reduces the amount of water that needs to be applied. Soil improvements result in better water retention and healthier plants. For example, adding 3 cubic yards of well-aged compost per 1,000 square feet of landscaped area will hold water and increase plant growth.

Mulches are materials applied to the soil. A three-inch layer of mulch will improve the appearance of landscaped beds, as well as reduce weed growth, moderate soil temperatures, minimize erosion, and help retain soil moisture. Mulched beds can replace turf areas, reducing water requirements.

The use of mulches and soil amendments can be promoted during a landscape irrigation audit, with the auditor offering information on where to obtain these materials and how to apply them. The utility may also offer rebates for the use of mulches and soil amendments. Some municipalities and local governments have codes requiring the application of these materials for new construction.

Workshops, Demonstration Landscapes, Requirements

Water-efficient landscaping can be promoted through workshops to educate both water customers and landscapers. It can also be promoted through the distribution of printed information about appropriate plants, soil care, and other water-efficient landscaping techniques. University Extension agents will be willing to assist in crafting and delivering this education.

Demonstration gardens are important, because builders, landscapers, and water customers need to see that this type of landscaping can be attractive. Location of the demonstration landscape should be prominent, such as in a well-visited park, as in Figure 4-9, or outside a city hall. A local landscape architect may be willing to partner with a small or medium-sized utility on design of a demonstration water-efficient landscaping garden, in return for having his or her name on the signage. Other funding partners should be sought for a greater community buy-in for the landscape.

The local native plant society may partner in planting and possibly even maintenance, if the garden features 100 percent natives or nearly so. University Extension offices usually have Master Gardeners, trained volunteers that assist extension agents in educating the public. These Master Gardeners may welcome the opportunity to be involved in the demonstration landscape; however, they will not be available to maintain it, as their mission is education. For best success, in terms of aesthetics and marketing of this type of landscape, professional design by a landscape architect is essential. Interaction with the landscape design community can be a vehicle to educate landscape architects on water efficient design principles and plant choices.

Before embarking on plans for a demonstration garden, a utility should ensure that it can dedicate staff time or has long-term dedicated volunteers to maintain it. A water-efficient landscape takes less maintenance than the typical turf landscape, but

Table 4-1. Available Efficient Technologies

Device Description	*Flow Rating	Estimated Cost ($ per unit)			*Device Life (yrs)	End Use Reduction		Status
		Supply	Install	*Annual		gal/ cap/day	%	
Bathroom								
Low Flow Showerhead	Water Sense labelled	$10-50	By user		5-10 yrs		21%[a]	Voluntary
Shower Flow Restrictor	2.5 gpm	$5	By user		5 yrs		21%[a]	Voluntary
Faucet Aerator	0.5-1.5 gpm	$1-3	By user		5 yrs	0.3[c]		Voluntary
Toilets								
Low flow toilets	1.6 gal/flush	$100-350+	$100-250		20-30 yrs		52%[a]	Required for New
New Fill Valve & Flapper/ Leak Repair	NA	$10	$25			5+		Voluntary
Water Displace-ment Device	0.5-0.7 gal/ flush	$2	$10		2 yrs	2-34		Voluntary
High Efficiency Toilets	1.28 gal/ flush	$100	$100-250		20-30 yrs		63%	Voluntary
Composting Toilets	0 gal	$2,000	$500	$200	20+ yrs	20.1[a]		Voluntary
Kitchen								
Faucet Aerator	2.2 gpm	$3	by user		5 yrs	0.3[c]		Required for New
Insulate Hot Water Pipes	NA	$25/100 ft.	by user		10 yrs	2d		Voluntary
Efficient Dishwasher	5.5 gal/load	$500-900	$200		10-15 yrs	0.5[e]		Voluntary
Laundry								
Faucet Aerator	2.2 gpm	$5-10	by user		5 yrs	0.3[c]		Required for New
Horizontal Axis Clothes Washer	15-25 gal/ load	$600-1000	$100		15-20 yrs		35%[a]	Voluntary
General Household								
Pressure Reducer	<80 psi	$90	$200-400		20+ yrs	3-6[d]		Regulated
Submeters on Apt Units	NA	$50	$200	varies	20+ yrs		15[f]	Voluntary
Household Leak Repair	NA	varies	varies		5 +/- yrs	5.0[a]		Voluntary
Graywater Systems	NA	$500 to $10,000	$200-500		10-20 yrs	10-50		Regulated

Source: From Maddaus et al. 2009 with permission.

Table 4-1. Available Efficient Technologies

Device Description	*Flow Rating	Estimated Cost ($ per unit)			*Device Life (yrs)	End Use Reduction		Status
		Supply	Install	*Annual		gal/ cap/day	%	
Public Education	NA			~$1-2/ person	1		1-5%[d]	Voluntary
Landscape Irrigation								
Drip Systems	NA	$15 for 20 plants	by user		10 yrs	Varies[g]		Voluntary
Micro-Spray Systems	NA	$25 for 20 sq.ft.	by user		10 yrs	Varies[g]		Voluntary
Landscape Irrigation (continued)								
Hose Timers	NA	$10-40	by user		5-10 yrs	Varies[g]		Voluntary
Rain Sensor	NA	$30	100		10 yrs		5-10%	Voluntary
Hose Shut Off Valves	NA	$3-8	by user		5-10 yrs	Varies[g]		Voluntary
Soil Moisture Sensors	NA	$30/valve	100		10 yrs	Varies[g]		Voluntary
ET Irrigation Controllers	NA	$500-1,000	$200-500		10-15 yrs		15-25%[h]	Voluntary
Native Plants	NA	Varies	Varies		5-20 yrs		15%	Voluntary
Mulch	NA	1/sq. ft.	by user		5 yrs	Varies[g]		Voluntary
Synthetic Turf	NA	$4/square foot	$3.50/ square foot		10-15 yrs		95%	Voluntary
General Commercial (Other than above measures)								
6-Liter (Commercial) Toilets	1.6 gal/flush	$300-450	$100-250		20+ yrs	5.7[n]		Required for New
1.0 gal/flush Urinals	1 gal/flush	$600	$100-400		20+ yrs	3i		Required for New
0.25--0.5 gal/ flush Urinals	0.25-0.5 gal/flush	$600	$100-400		20+ yrs	4.5[i]		Voluntary
Restaurant Pre-rinse Spray Valves	1.6 gpm	$70	$120		10-15 yrs		50%[l]	Required for new or replacement
Digital X-Ray Machines	No water	$5,000-35,000	varies		20+ yrs	Varies		Voluntary
Large Efficient Commercial Dishwashers	Varies	$30,000-60,000	$500		20+ yrs	Varies		Voluntary
Sensors for Steam Sterilizers	Low water use	$2,500	$100-400		20+ yrs	Varies		Voluntary
Commercial Food Steamers	Varies	$5,000-15,000	$500		10-15 yrs	Varies		Voluntary

Source: From Maddaus et al. 2009 with permission.

Table 4-1. Available Efficient Technologies

Device Description	*Flow Rating	Estimated Cost ($ per unit)			*Device Life (yrs)	End Use Reduction		Status
		Supply	Install	*Annual		gal/ cap/day	%	
Air Cooled Ice Machines	Varies	$4,000 - 12,000	$500			Varies		Voluntary
Commercial Clothes Washers	Varies	$1,600	varies		20+ yrs	Varies		Voluntary
Commercial Laundry Recycling Systems	NA	$100,000 and up	included		20+ yrs		65%[o]	Voluntary
Automatic Meter Reading Systems	Varies	$200+/ home when done on large scale	included		20+ yrs	Varies		Utility Measure

Source: From Maddaus et al. 2009 with permission.

Notes: Demand reductions shown are preliminary and subject to change. Actual savings vary with household size, current devices or technology in use, portion of water used in the landscape etc.

* Denotes *Where Applicable*, water used may vary depending upon water pressure and maintenance.

a Nelson, J.O., "Household End Uses of Water", posted on www.waterwiser.org, 1999

b Based on retrofit to the equivalent of a new low flow showerhead (see note a above)

c Total savings if installed on kitchen and bathroom sinks, based on note a above

d Maddaus, W.O. "Water Conservation", American Water Works Association, 1987.

e Based on Mayer, P.W., et al "Residential End Uses of Water, American Water Works Association Research Foundation", 1999, and an assumed water savings of an efficient machine of 5 gallons per load.

f Dietemann, A. "Sub-Metering: The Next Big Frontier?" Seattle Public Utility, Conserve99 (AWWA), February 1999, Orlando, Florida.

g Depends on amount of water used outside and interaction with other outdoor measures

h Berg, J.O. et al. "Residential Weather-Based Irrigation Scheduling: Evidence from the Irvine "ET Controller" Study, June 2001, Irvine Ranch Water District, Calif.

i Compared to 2.0 gpf urinals using three flushes per employee per day

j Compared to 1.0 gpf urinals using three flushes per employee per day

k Assuming elimination of one extra 1.0 gal flush per employee per day

l Personal Communication with John Koeller, Koeller and Company, June 2002.

n Compared to 3.5 gpf toilets using three flushes per employee per day

o Personal Communication with Randall Jones, Wastewater Resources, Inc., Scottsdale, Ariz., February 2000.

some weeding will be required over the life of the garden. Many demonstration gardens have become unsightly for lack of maintenance. The landscape will also need to have mulch replenished at least annually, and funds should be budgeted accordingly.

Signage should be planned from the outset, as in Figure 4-9, as the purpose of the garden is to teach identification of the plants and about water-efficient landscaping principles. The signage can list businesses and entities that have contributed to funding. The garden can be used as a site for classes, which may be taught by master gardeners or other plant experts.

Utilities can offer service connection fee discounts to builders if they install water-efficient landscaping, with the utility clearly articulating what is acceptable. Southern Nevada Water Authority has had success with its Water Smart Home program

Figure 4-9. Demonstration water-wise landscape in city park includes informative signage, explaining the purpose of the project, and small signs identifying individual plants, Edgewater, Fla. (Photo: Water Media Services)

in providing a marketing advantage to builders who become a part of the program and follow program requirements. A landscape ordinance, as in Landscape Irrigation Ordinances below, can also require the use of low water-use plants for new landscaped areas.

Programs carried out in warm, dry climates demonstrate that the potential for water-use savings from water-efficient landscaping is high, ranging from 40 to 60 percent (Maddaus 1987). The Southern Nevada Water Authority in Las Vegas assists customers to convert from turf-dominated landscapes to water-efficient landscaping and found average annual water savings of 30 percent with their program (Sovocool et al. 2004, Sovocool 2005).

IRRIGATION SYSTEMS

Most new homes come with automatic irrigation systems, and in some parts of the country widespread use of residential automatic irrigation systems dates back to the 1980s. Unlike toilets and other indoor fixtures that operate simply and reliably, irrigation systems are complex and have a vast capacity for water wastage. All phases of irrigation have potential for water efficiency improvements: design, installation, and maintenance.

Programs to improve irrigation efficiency are challenging but have great potential for water savings. Finding qualified irrigation contractors and forming partnerships with them can assist in success of these programs. As outdoor water use increases and causes peaking supply problems, landscape conservation programs are a very cost-effective means of reducing this discretionary demand (Vickers 2008).

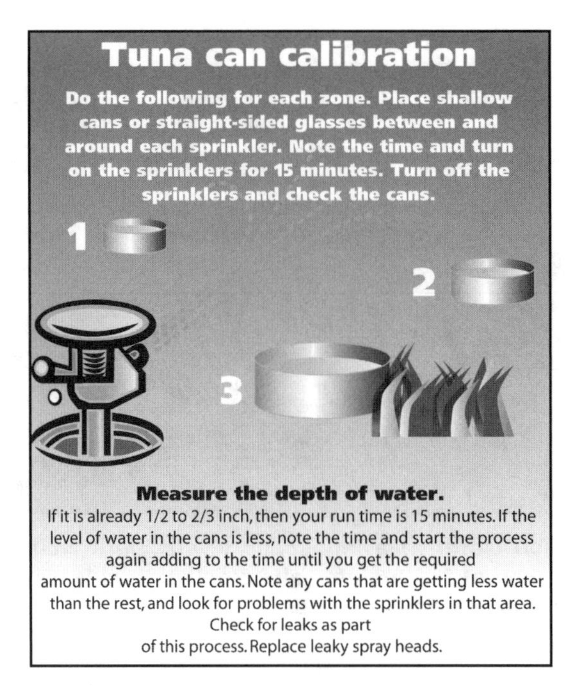

Figure 4-10. A typical lawn-watering guide showing the customer how to check distribution uniformity and determine how long to run the irrigation system to put down the required volume of water. (Graphic: Water Media Services)

Lawn-watering Guides

Printed lawn-watering guides educate homeowners about the actual water needs of irrigated turf grass and discourage overwatering. These guides typically emphasize the different water needs of turf grass in different seasons. They also instruct the homeowner how to measure the distribution uniformity of the irrigation system (for example, with tuna or cat food cans), as in Figure 4-10, and how to calculate the time required for watering. A lawn-watering guide should have information applicable for in-ground irrigation systems as well as moveable hoses and sprinklers.

A small or medium-sized utility can receive assistance in developing appropriate information for a lawn-watering guide from local University Extension agents specializing in soils and irrigation. Again, a utility should establish a relationship with reputable certified irrigation contractors, and these professionals may be utilized to check the appropriateness of consumer messaging. Lawn watering guides can be printed by the utility and also submitted to newspapers and featured on local government websites.

Watering Restrictions

Watering restrictions have been used by utilities around the country to reduce overwatering. Most prohibit watering during hours of high evaporation. Many specify even and odd watering days based on home addresses. Watering hours are widely advertised in print media and on the radio, particularly during drought periods.

Publicity on watering restrictions, however, is not enough. Hillsborough County, Fla., has had an aggressive program of watering restrictions enforcement, which it continues long term, because areas of enforced restrictions have significant demand reduction compared to areas without enforced restrictions. Documenting specific demand reduction from watering restrictions, and distinguishing it from the savings of other programs such as reclaimed water, toilet rebates, showerheads, and public awareness, however, is difficult. Even with aggressive enforcement, violations continue (Davis 2009). Collected fines from watering restrictions can be used to fund other water conservation programs, as in Chapter 3, Funding.

Small and medium-sized utilities often designate meter readers and other utility workers to conduct watering restrictions enforcement. Some cities have used local police or code enforcement officers to ensure that watering restriction enforcement is a priority. City managers or commissioners must create this directive.

Certified Irrigation Contractors

Small and medium-sized utilities can develop lists of certified contractors for referrals to customers. These same contractors might be employed in an irrigation audit program. See the Irrigation Association and WaterSense websites for certified irrigation contractors in the utility area.

Landscape Water Audits, Residential

Water audits involve identifying water uses and discussing water-use practices with the customer. They are typically conducted at the request of the customer, who has responded to the utility's offer of a free water audit. To be most cost-effective, they should target the highest seasonal customers, regardless of residential, commercial, or institutional status. The audits should be advertised as available to customers during spring and summer. The City of Austin established a free irrigation audit program in the 1980s to combat the dramatic spike in water demand during the summer months due to landscape irrigation. Commercial, multifamily or residential customers may request an evaluation, and receive rebates for installing recommended upgrades, such as pressure reducing valves or new controllers. The city has conducted over 8,500 irrigation audits and attributes 1.15 million gpd saved to this effort (Gregg 2006).

Either utility staff members or contractors will perform the audits. Training to carry out an irrigation system audit is more involved than that to carry out an indoor audit, and in areas with extensive use of automatic irrigation systems, employing a trained irrigation auditor to perform the audits is helpful. These individuals will be familiar with the wide variety of older irrigation equipment that is currently in place in the area. The Irrigation Association, www.irrigation.org, trains and certifies auditors. A state may have an irrigation association that provides similar training. WaterSense also certifies landscape irrigation professionals.

In the water audit, an irrigation schedule for the lawn is usually provided along with information on water-saving practices for outdoor water use. The utility may offer to install rain sensors, microrotors, or other irrigation system upgrades as part of the audit. Customers are attracted to free items, and if interest in the audit is not extensive, the utility should publicize "freebies." Indoor retrofit items can also be included.

A suggested procedure for a brief residential landscape and irrigation audit is provided in Appendix B. A full irrigation audit with catch-can tests to determine distribution uniformity of each zone is appropriate for a large commercial landscape audit,

Figure 4-11. Irrigation Audit offer in which customer pays $25 and utility pays contractor $25 or remainder of going rate for a one hour audit. Several certified irrigation contractors/ auditors can be on a referral list for this program. (Graphic: Water Media Services)

but is generally not cost-effective at the residential level. Typically, if a full audit is carried out, fundamental design problems are identified, such as zones containing sprays and rotors, or turf and low water use landscape plants on the same zone. Retrofitting to improve efficiency can be expensive for the customer (although new technologies such as retrofit with microrotators can reduce the costs).

Some utilities have offered cost-shares for retrofits following audits, specifying a list of acceptable retrofits. System efficiency can be greatly improved by these retrofits, and for research and verification purposes only, the utility can carry out before and after assessments of distribution uniformity through catch can tests.

Small and medium-sized utilities can most productively focus on timer settings (described in the following section) if they develop an audit program. The entire water audit process will take an hour or more, depending on the type of irrigation system and detail of auditing performed. Costs for the utility include staff or contractor time, appointment scheduling, follow-up, and materials and conservation devices. If the marketing effort makes clear the water and wastewater bill reduction that a residential customer will achieve following the audit, customers might consider paying part of the cost of an audit, for example $25 as in Figure 4-11. A utility offer of water audits can be promoted through public information campaigns and media coverage.

Timers should be able to be set in increments of minutes. Some earlier models could only be set in increments of 15 minutes, 30 minutes, 45 minutes, or an hour. If the customer calibrates the sprinkler system for each zone to calculate run times, as explained in the lawn watering guides (Figure 4-10), the highest efficiency can be achieved. Alternatively, a trained irrigation contractor can estimate run times for zones based on the application rates of the irrigation equipment in that zone. For example, to put down ½ inch in sandy soil, spray zones are typically run 15-20 minutes and rotor zones 35-40 minutes. Local extension service can provide recommendations for local soil conditions.

Irrigation Controllers

Irrigation controllers (also called *timers or clocks*) for automatic irrigation systems have become increasingly sophisticated. However, earlier models, of which there may be thousands installed in a utility service area, have various problems. Earlier timers may revert to a default setting of watering every day following a power outage. Modern timers have back up batteries and nonvolatile memory. Modern timers are generally much easier to set and adjust, although the presence of multiple start times can confuse users.

The SMART Water Application Technology (SWAT) program of the Irrigation Association and the USEPA WaterSense Program carry out third party performance testing for irrigation system add-ons like soil moisture sensors and rain sensors and may soon evaluate timers. Until this information is available, the professional judgment of reputable irrigation contractors can be followed regarding best performing models. In general, irrigation supply stores are better sources of professional grade equipment than home supply stores and have staff able to assist customers.

JULY
SMART
IRRIGATION
MONTH

The SMART Water Application Technologies is a national partnership initiative of water purveyors and irrigation industry representatives through the Irrigation Association (www.irrigation.org). In the same way that the performance testing of toilets has propelled improvements by toilet manufacturers, development of testing protocols and testing of irrigation equipment will propel improvements for irrigation equipment. Products that have been SWAT tested have met set criteria for performance.

Unrelated to the performance of timers is lack of knowledge of how to operate them by the homeowner. Many customers that purchase a home with an existing in-ground irrigation system do not know how to use the timer. In a study of water reduction potential of a landscape ordinance in Volusia County, Florida, billing records for metered irrigation accounts, most of them using reclaimed water, were examined over a year-long period before and after enactment of the ordinance. Results indicated that water use did not decrease in the months after the installation of the landscape. The same level of irrigation that had been set by the contractor to establish the landscape frequently remains in place (Green 2006). A program to assist homeowners in timer resetting after establishment could reduce water use for those homes by more than 50 percent.

Knowledge of how to adjust watering times seasonally is also important. In many parts of the country, timers are turned off entirely in the winter. However, in warmer regions where timers are not customarily turned off, propelling customers to reduce the amount applied in winter is important. The customer can reduce the run time by 20 percent or 50 percent, using the water budget feature of more advanced timers. Another strategy is to reduce the days by half. The Southwest Florida Water

Figure 4-12. Variety of older timers turned in during pilot distribution of SMART controllers, www.socalwatersmart.com (Photo: Metropolitan Water District of Southern California)

Management District has a Skip-A-Week campaign for winter watering (See the model press release on this campaign in Appendix C).

Small and medium-sized utilities in areas where residential irrigation is a large part of use should consider programs educating customers on timer settings, beyond just the provision of lawn-watering guides. Workshops can be offered. Irrigation contractors are best able to assist with timer resetting workshops, because they are familiar with the timer models that have been sold over the years. The Irrigation Association's SWAT program has developed a July Smart Irrigation Month program that can be the occasion for a series of workshops. For ideas, visit the Smart Irrigation Month website www.irrigation.org/sim (Figure 4-12).

Adjusting timer settings is generally a part of a residential landscape audit. However, the auditor should not just set the timer for the customer but rather make sure the customer knows how to operate the timer and understands correct settings for different seasons. Be aware that most auditors will not teach the customer without being specifically requested to do so.

Requirement for the builder to provide the timer manual may be part of a landscape ordinance. Timer manuals can be downloaded from manufacturer websites for many older timers, and the utility can provide links to the websites of manufacturers of the timers commonly used in the area.

As previously mentioned, alliances with reputable irrigation contractors/auditors can assist in development and implementation of all irrigation programs. Coordinating with programs developed by other utilities in the same area can be beneficial.

Timer Add-ons: Rain Sensors

In states where rainfall occurs in the months of irrigation (for example, summer months as in the Southeast), rain sensor devices are important add-ons to automatic systems. Rain sensor devices (also called rain switches or rain shut off devices) turn off the automatic irrigation system when a set amount of rainfall has been received and

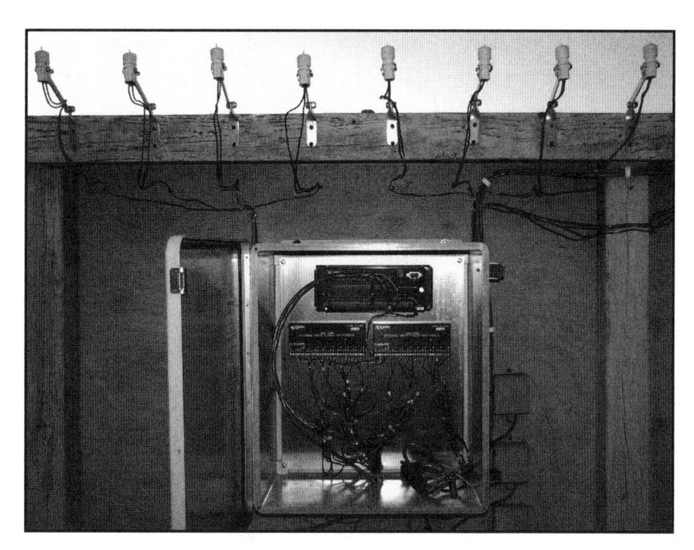

Figure 4-13. Rain sensors being tested for performance at the University of Florida. (Photo: Bernard Cardenas-Lailhacar)

allow the system to resume its preset schedule when the device dries out. These add-ons are inexpensive, selling at irrigation supply stores for about $20, with irrigation contractors paying even less. Typical installation time is about an hour for a contractor. New wireless rain sensors take even less time to install. Cost of wireless models averages about $50 not including labor.

Installation of rain sensor devices has been required in Florida on all new automatic irrigation systems since 1991. The states of Georgia, Illinois, Minnesota, New Jersey, and Connecticut plus several cities and counties, mostly in the Southeast, have also enacted rain sensor requirements. Rain sensor devices have much more limited usefulness in western states and parts of the country where the rainy season is in winter, when irrigation may be minimal or nonexistent. It should be noted that enforcement of a rain sensor ordinance through inspection is necessary to be sure that these add-ons are installed and functioning. In Florida, the rain sensor requirement, however, is part of Florida water law, but not in Building Code, other than as an appendix that is only optionally adopted by counties. Consequently building inspectors have traditionally not inspected for these devices (Green 2008).

If the utility is in an area with summer rainfall, if the customer base is mainly residential, and if the utility's water demand does NOT drop in the rainy season, the utility should consider a rain sensor installation program. Rain sensors can be purchased in bulk through a local irrigation store and given away at water efficiency workshops, taught by a University Extension agent. Alternatively, the utility can cost-share with customers the cost of the device and installation through a rebate program.

Previously, rain sensor devices had a reputation for not working. Much of this involves incorrect installation, including issues of compatibility of rain sensors purchased at home supply stores with timers by different manufacturers. Providing installation along with any giveaway can be beneficial. Performance and reliability of rain sensor devices have improved and will continue to improve, since rain sensors are included as SMART controllers and are being tested under the SWAT program. Bench testing of rain sensors by University of Florida researchers is shown in Figure 4-13.

Small and medium-sized utilities in parts of the US requiring rain sensors can ensure that the statute is enforced and can consider giveaway programs, alone or with installation, as a means to revive interest in the devices. An ordinance at the municipal level can require the devices to be installed on all systems, even those grandfathered in before the effective date of a state requirement. Estimated savings are 15-25 percent of annual irrigation use in areas with summer rainfall (Cardenas-Lailhacar and Dukes 2007).

Other Advanced Irrigation Equipment Promoted by Utilities

Heads with pressure regulation prevent water loss through misting and greatly increase distribution uniformity. Check valves are important add-ons for low lying areas, to prevent irrigation water from running off the lot. Requirement for these two types of equipment can be part of a municipal landscape ordinance for new installations or major renovations.

Microrotors are a type of rotor with multiple rotating streams of water (Figure 4-14). They are the size of spray sprinkler nozzles and fit on standard spray sprinkler bodies. Multistream multitrajectory matched precipitation rate (MSMT-MPR) is the generic term for these devices; however, an Internet search under this generic name may bring up little information in comparison to a search under the name MP Rotator (Hunter Industries' brand name following purchase of the device from Walla Walla Sprinkler Company). Other companies are perfecting versions of the devices, and the SWAT program plans to test these devices, so the generic name or the shortened name microrotor may become more widely used.

The benefits of microrotors in improving system efficiency are two-fold. First, a large contributor to inefficiency of in-ground automatic irrigation systems is the mixing of two types of equipment (rotors and sprays) on the same zone. It is expensive in both equipment and contractor time to retrofit mixed zones to entirely sprays or entirely rotors. However, microrotors can be substituted for sprays in a mixed zone. By matching the precipitation rates, microrotors will greatly improve the distribution uniformity (DU), a key measure of irrigation system efficiency. This benefit occurs in irrigation systems throughout the country and is of reasonable cost compared to a major zone retrofit.

The second major benefit of microrotors occurs only in clay soil areas or other areas where soils have low permeability. To avoid runoff in these areas, irrigation water must be applied slowly. One way to achieve slow application is through multiple start times, in other words, applying water for a short time and then applying more after that water has soaked in. The latter technique requires a sophisticated timer and some skill in programming. However, the microrotor applies water so slowly that it achieves the same objective.

Many large utilities give rebates for landscape renovations that include these spray head substitutes. Among the leaders in promoting these devices have been Metropolitan Water District of Southern California, Irvine Ranch Water District, Municipal Water District of Orange County, San Diego County Water Authority, East Bay Municipal Utility District, Central Utah Water Conservancy, Denver Water, Aurora

Figure 4-14. Microrotors put down multiple rotating streams of water. (Photo: Hunter Industries)

Water (Colorado), Colorado Springs Utilities, and Seattle Public Utilities. Websites for these utilities are listed in Appendix A.

A small and medium-sized utility with high residential irrigation water use might follow the lead of larger utilities in assisting customers to install these devices through rebates. A sound precursor to embarking on such a program is to establish a relationship with several reputable and certified irrigation contractors. If someone within the utility has irrigation installation skills, this person can be sent to classes offered by the Irrigation Association or state irrigation groups. The SWAT website (www.irrigation.org/SWAT/Industry) will have information on whether microrotors may have been tested, which will make recommendation of specific brands easier. Rebates for microrotors can be combined with the irrigation audit program and may make the audit program more attractive to residents.

Microirrigation

Microirrigation (sometimes written as micro-irrigation) or low volume irrigation is effectively used as part of water-efficient landscaping. Microirrigation is often referred to as the earliest known type, which is drip irrigation. Other types of microirrigation are microsprays and bubblers. A residential microirrigation system may include misters/sprayers/minisprinklers, drip lines, PVC pipe, automatic controller, valves and pressure regulator, filters, and fertilizer injectors, as in Figure 4-15. The size and type of the plantings, as well as soil type, determine microirrigation system installation.

Microirrigation allows the transfer of water directly to the roots of plants needing watering, reducing evaporation. Weed growth, erosion, and water wastage can be reduced. Conventional irrigation equipment (sprays and rotors) apply water in gpm; however, microirrigation applies water in gallons per hour (gph) Bubblers may apply at rates of less than 60 gph, microsprays at less than 45 gph, and drip emitters at less than 2 gph (Haman et al. undated).

In-line drip tubing is generally half inch in diameter and has emitters spaced 12 in. or 18 in. apart. It may be wrapped around a tree to provide adequate water or can be snaked through vegetation beds or laid out in a grid. It is generally not used to irrigate turf. In a residential or commercial setting, drip tubing is usually buried under mulch.

In-line drip tubing is easy to install by a contractor. One quarter inch tubing and other emitter parts can be affixed in the in-line tubing. In high quality drip tubing, as shown in Figure 4-16, the emitters are pressure-regulated. Root inhibitors in the plastic prevent roots from growing into these emitters, and they are resist clogging. The microirrigation system requires very good filtration (most often recommended is 200 mesh screen) even with a good quality water supply.

In-line drip tubing works better in heavier soils. There the plume of water spreading from the emitter moves a smaller distance. In highly porous sandy soils, the water may drain downward and out of the plant root zone quickly. Soil amendments and mulch may help this loss of water to the root zone.

Microsprays, as shown in Figure 4-17, are small spray devices mounted on short risers connected to ¼-in. flexible tubing connected to ½-in. poly tubing main lines. The emitter sprays covering individual plants or a small group of plants. Microsprays are appropriate for landscape beds, trees, and other nonturf areas. It should be noted that microsprays are not considered effective in windy, arid areas, as the water mists and frequently does not saturate the soil. However, in Florida, where microsprays were first developed for the citrus industry, they perform well. Installed by an irrigation contractor, they can be connected to an automatic irrigation system. A homeowner can also install a microspray system, using inexpensive widely sold kits that are attached directly to the hose bib.

Benefits of microsprays over inline drip tubing are that the emitter can be seen, and clogged emitters can be easily cleaned or replaced. In contrast to in-line drip tubing, microsprays apply the water directly to target plants, rather than emitting water every 12 or 18 in., regardless of whether there is a plant at that point.

Drought tolerant plants will need no water after establishment and the microirrigation system can be considered a temporary system, as in Figure 4-18. It can be left in place and turned on in times of drought or even taken out and components used elsewhere after a period of a few years. Microirrigation can be used in conjunction with an automatic irrigation system or in a temporary system, with or without a timer.

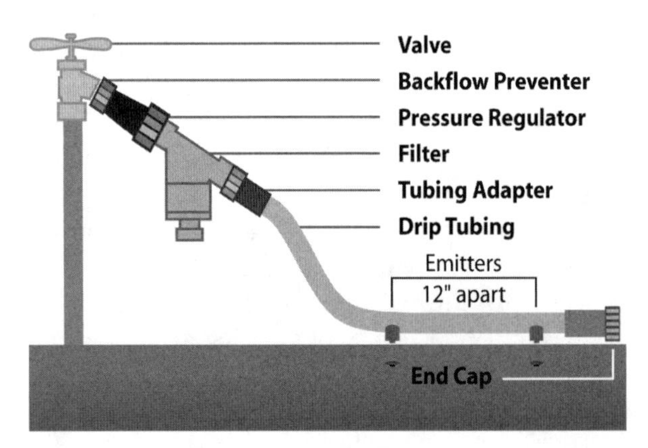

Figure 4-15. Hose bib connection of in-line drip tubing. In-line drip tubing installed by a contractor is connected to PVC main lines as part of an in-ground irrigation system. (Graphic: Water Media Services)

Figure 4-16. In-line drip tubing for sale in irrigation store in Volusia County Florida, following passage of an ordinance limiting high volume irrigation to no more than 60% of the landscape. (Photo: Water Media Services)

Figure 4-17. Microsprays allow water to be directed to individual plants. Shown is Mister Landscaper emitter, part of a homeowner-installed kit. (Photo: Mister Landscape Inc.)

Figure 4-18. A timer attached to a temporary microirrigation system to landscape beds. Drip lines will be taken out after establishment. (Photo: Water Media Services)

Watering restrictions typically do not apply to microirrigation, and some customers may run these zones for long periods of time, which may negate some of the water savings. However, because microirrigation is not used to irrigate turf, utilities promoting it for landscape beds find it is generally water conserving. Vickers (2001) cites 15–40 percent use reduction over typical irrigation systems, while other sources estimate savings as high as 70 percent (Southwest Florida Water Management District 2006).

Small and medium-sized utilities can follow the lead of large utilities in the area by promoting use of microirrigation. A typical way is by limiting percentage of a new or renovated landscape that can be high volume, which is often defined as emitting water in gallons per minute (microirrigation emits water in gallons per hour). Installation of a microirrigation system can also be encouraged following residential water audits.

Swimming Pools

There are 9 million pools in the United States, all but three percent of them residential. Swimming pools were found to use approximately the same amount of water on a square footage basis as irrigated lawn area, using water use estimates for a typical home in Sacramento and Tampa (Horner 2009, Maddaus and Mayer 2001).

Two major sources of water loss in pools are evaporation and filters and backwashing. If a pool is heated, as much as 70 percent of heat is lost through evaporation. In the summer, evaporation ranges from five to ten inches a month, according to the U.S. Department of Energy (Horner 2009).

Using a pool cover when the pool is not in use, as in Figure 4-19, reduces heating loss by 50–70 percent. Pool covers also reduce the amount of make-up water required by 30–50 percent, and reduce chemical consumption by 35–60 percent according to the U.S. Department of Energy (Horner 2009).

A small or medium-sized utility in an area with large numbers of residential pools can publicize water- and energy-saving benefits of pool covers to target customers.

Figure 4-19. Presence and use of a pool cover is a major factor in reducing water loss in a pool. (Photo: Water Management Inc.)

PROMOTIONS AND INCENTIVES

Promotions (Information-based Incentives)

Many of the conservation measures previously described result in reduced water, wastewater, and energy bills for consumers and this information can be provided to customers to propel voluntary retrofits. For example, if a family has a 3.5-gpf toilet, four people in the household with an average flush rate of five times per day per person, by changing to a WaterSense labeled 1.28-gpf toilet, this family can save over 16,000 gallons per year (3.5–1.28 savings × 4 people in family × 5 flushes per day × 365 days per year). A utility can then use water rate per gallon saved as well as any charges related to wastewater to get a savings rate per year for the customer. This information can advise customers whether the payback period on replacing a plumbing fixture or appliance is short enough to warrant an investment on their part. The WaterSense website contains a payback calculator at www.epa.gov/watersense.

A utility joining the USEPA WaterSense program as a promotional partner (at no cost) can use the WaterSense logo to promote WaterSense certified products. Up-to-date information, accessible to customers, is available on the WaterSense website. By informing customers about the payback of making certain changes—either by replacing fixtures, or changing behavior—some level of savings can be achieved.

Financial Incentives

Connection fee discounts or customer rebates can be offered by utilities for properties that meet certain conservation criteria. Possible eligible practices include

- Installation of high efficiency plumbing fixtures, for example those labeled by WaterSense.
- Limitations on turf areas and thus irrigation for landscaping and use of water-saving groundcovers and plants.
- Design of efficient landscape irrigation systems.
- Soil preparation to reduce water needs of turf and plants.
- Sustainable sites initiatives.
- Use of multiple-pass water for heating, ventilation, and air conditioning (HVAC).
- Submetering of multi-family dwelling units.

For new home construction, some builders are interested in showcasing their new homes as "green" and may go through one of the national or regional "green build" certification programs, such as the United States Green Building Council's LEED for Homes program, National Association of Home Builders Green Home standards, or regional programs. These programs award credits for water efficiency measures. Detailed standards may be listed, or meeting WaterSense standards for fixtures may be required to earn credits. The USEPA now has a WaterSense Water-Efficient Single-Family New Homes standard.

A small or medium-sized utility in a high growth area might give discounts on water hook-up fees or fast-track permitting for builders that follow the WaterSense Water Efficient Single Family New Home standard or one of the green build standard programs.

REGULATIONS, WATER WASTAGE, AND RETROFIT ON RESALE

Prohibition of Water Waste

Utilities may prohibit water waste through resolutions and ordinances or operating rules, if they are not part of a municipal or regional government. Typical ordinance language regarding water waste (modified from Southern Nevada Water Authority www.snwa.com and Sunnyslope County Water District, Hollister, California www.sscwd.org) prohibits:

- Indiscriminate or excessive water use, which results in water waste.
- Any water that sprays or flows off a property.
- Use of sprinklers during hours of high evaporation.
- Use of potable water to irrigate turf, lawns, gardens, or ornamental landscaping (between 9:00 a.m. and 5:00 p.m.) by means other than drip irrigation or hand watering with a quick-acting, positive shut-off nozzle.
- Allowing water waste from easily correctable leaks, breaks, or malfunctions, after a reasonable time within which to correct the problem (Failure to repair a malfunction in an irrigation system or supply line within 48 hours).
- Failure to discharge swimming pool or spa drainage into a public sanitary sewer, if available.
- Washing of cars, buildings, or exterior surfaces without the use of a quick-acting, positive shut-off nozzle.
- Use of potable water to wash sidewalks or roadways, when the use of air blowers or sweeping would provide a reasonable alternative.

Some utilities (for example, the Las Vegas Valley Water District) offer a *Curbing Water Waste* class throughout the year to help customers identify and correct water waste. The ordinance can specify a fine for violation with a waiving of the fine if the resident attends a specific workshop.

Other water waste restrictions can be included in a drought ordinance and enforced when a predetermined water shortage indicator level is reached. These include

- Failure to comply with drought restrictions.
- Operation of decorative fountains, even if they use recirculating systems.
- Use of water for construction purposes, such as consolidation of backfill, except when no other method can be used.
- Hydrant flushing, except where required for public health and safety.
- Refilling existing private pools, except to maintain water levels.
- Restaurant water service unless on customer request.

Plumbing Codes and Retrofit on Resale

Local governments can enact more stringent requirements than federal and state codes for plumbing fixtures in new construction. Regulations can require the use of water-saving devices, such as high efficiency toilets and showerheads. If a local government ordinance requires installation of WaterSense labeled products in all new homes, they will be securing fixture volumes 20 percent below current code (Pape 2008).

To bring older properties up to current federal codes or more stringent local codes, local governments are increasingly utilizing retrofit-on-resale ordinances. Particularly in areas with slow growth or during slow growth periods, the potential to retrofit existing homes and businesses is great.

Local governments that have adopted ordinances requiring homes be retrofitted with water efficient fixtures on resale are City of San Diego, City of Los Angeles, City of San Francisco, City of Santa Monica, County of Santa Cruz, Monterey Peninsula Water Management District, and North Marin Water District. DeKalb County, Georgia has put in place an ordinance requiring retrofit to current code on hookup to the water system. This allows the utility to control the retrofit implementation.

A small or medium-sized utility can follow larger utilities in the area, or if political backing for a retrofit on resale ordinance is present in a community, the ordinances of the previously listed local governments can be researched (websites in Appendix A).

Landscape Irrigation Ordinances

Development and implementation of a landscape irrigation ordinance may reduce peak water demand. Ordinances usually require efficient irrigation systems and reduced areas of high volume irrigation, rather than focusing on requiring a specific list of drought-tolerant plants. There is always the possibility for the customer to overwater drought-tolerant plants. These ordinances can be specified to apply to new or also renovated landscapes of a specific size.

Considerable stakeholder input will be required to craft an ordinance that will be politically acceptable. Education of contractors, landscape architects, and code-enforcement officers is necessary. Also, enforcement of the ordinance must be set up and maintained for it to be successful.

One of the provisions of many landscape ordinances is a requirement for the irrigation contractor installing the landscape to provide "as-built plans" that can assist the customer in management of the irrigation system in the future. The as-built is a simple schematic identifying the irrigation zones plus the location of control equipment, as in Figure 4-20. An irrigation schedule for establishment and later maintenance, plus the timer manual, should also be provided. A basic requirement of a sound landscape ordinance is that turf and landscape beds should be on separate zones, referred to in the irrigation industry as *hydrozones*.

As an interim step before enacting (and enforcing a landscape ordinance), utilities can offer builders service connection-fee discounts for following the Water Sense Water Efficient Single Family New Home standard or one of the green build standard programs, as mentioned in the section on Financial Incentives. While the indoor portion of national standards easily applies throughout the country, the landscape portion of the standards needs to be regionally appropriate. University Extension staff can be helpful in checking information.

In California, all cities and counties must have a landscape ordinance in place; if they do not, the state's model ordinance applies. California uses evapotranspiration (ET), a measure of the amount of water required to maximize plant growth calculated from climatic conditions and factors such as temperature, solar radiation, humidity, wind, time of year, and precipitation. The general premise behind the ordinance is that the landscape should only be allowed to use 70 percent of the evapotranspiration requirements for turfgrass. California Law AB 2717, effective January 2010, amends earlier requirements and can be accessed at www.water.ca.gov/waterusefficiency. The Lousiana Green Law website is another source for examples of ordinances (Abbey 2009).

Small and medium-sized utilities can benefit from the experience of larger utilities in the same region regarding Landscape Irrigation Ordinances and should not embark on the process of developing an ordinance without establishing a sound plan for education of contractors and enforcement.

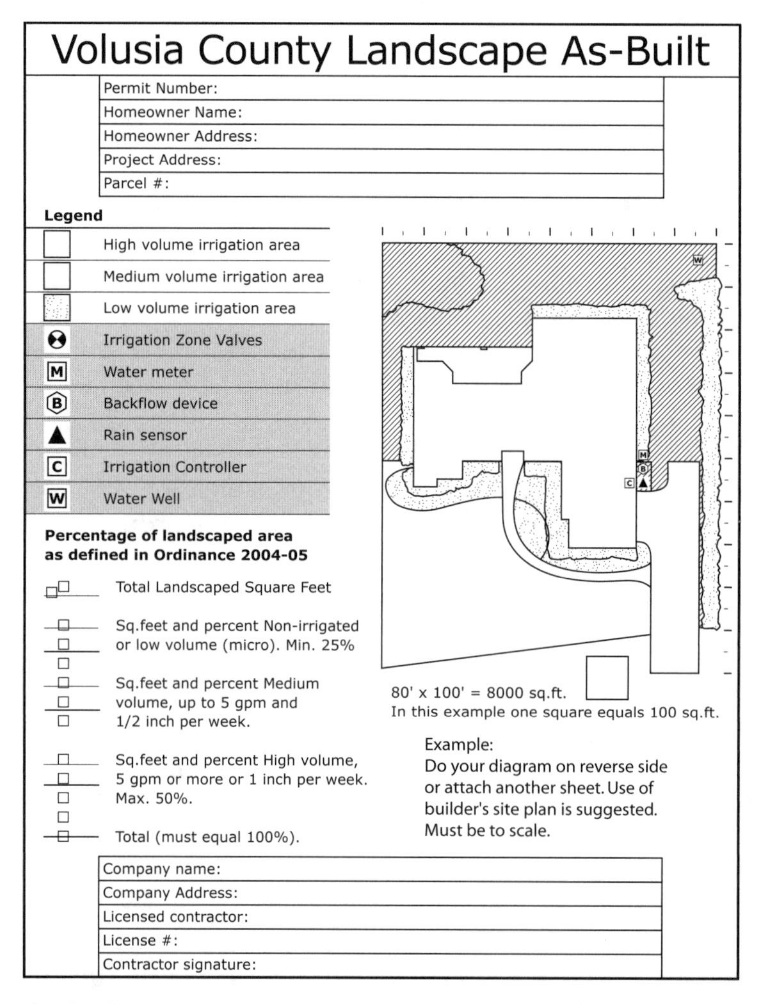

Volusia County Landscape As-Built

Permit Number:
Homeowner Name:
Homeowner Address:
Project Address:
Parcel #:

Legend

High volume irrigation area
Medium volume irrigation area
Low volume irrigation area
Irrigation Zone Valves
Water meter
Backflow device
Rain sensor
Irrigation Controller
Water Well

Percentage of landscaped area as defined in Ordinance 2004-05

Total Landscaped Square Feet
Sq.feet and percent Non-irrigated or low volume (micro). Min. 25%
Sq.feet and percent Medium volume, up to 5 gpm and 1/2 inch per week.
Sq.feet and percent High volume, 5 gpm or more or 1 inch per week. Max. 50%.
Total (must equal 100%).

80' x 100' = 8000 sq.ft.
In this example one square equals 100 sq.ft.

Example:
Do your diagram on reverse side or attach another sheet. Use of builder's site plan is suggested. Must be to scale.

Company name:
Company Address:
Licensed contractor:
License #:
Contractor signature:

Figure 4-20. As-built diagram to be filled in by irrigation contractor prior to securing permit for irrigation system. Homeowner is provided a copy, along with the timer manual and maintenance checklist, to improve ongoing system operation and maintenance. (Graphic: Water Media Services)

ALTERNATIVE WATER SOURCES

In addition to treated wastewater, several alternative sources are taking pressure off potable sources and providing important new resources. All of these except rainwater contain contaminants that have to be considered (Hoffman 2009).

Reverse osmosis and nanofiltration eject water may contain high salt content. Storm water and foundation drain water may contain pesticides and fertilizers. Air conditioner condensate may contain cooper when the coil has been cleaned. Pool filter backwash contains pool treatment chemicals. Cooling tower blowdown water contains cooling tower treatment chemicals. Gray water contains detergents and bleach. On-site wastewater treatment may contain human waste. All are under active research, as potable sources become scarcer (Hoffman 2009). A utility should research state codes and local regulations related to these sources.

Graywater

Graywater (sometimes written as greywater) derives from the bath, shower, washing machine, and bathroom sink. Blackwater is water flushed from toilets, as well as the kitchen sink, garbage disposal, and dishwasher. The latter has high concentrations of organic waste and cannot be used without extensive treatment.

Untreated graywater can supply most, if not all the irrigation needs of a residential landscape in a semiarid region. It may contain detergents with nitrogen or phosphorus, which are plant nutrients, but also sodium and chloride, which can be harmful to some sensitive species. The State of Arizona has been most active in allowing graywater use. The Water Conservation Alliance of Southern Arizona (Water CASA) has developed a well-illustrated publication *Graywater Guidelines*, that is downloadable from www.watercasa.org (Little 2001). The California Urban Water Conservation Council has also been active in promoting graywater technology. See www.cuwcc.org.

Sophisticated commercially available systems now treat graywater prior to disposal to reduce groundwater contamination and can remove pollutants and bacteria. The better systems include settling tanks and sand filters (Gelt 2009). With current costs of water low in most areas and costs of these graywater treatment technologies high, the systems are not yet cost-effective (Gauley 2008); however, in Green Building programs, credits can be earned through substituting graywater for potable water for irrigation or toilet flushing. As conventional alternatives become more costly and costs of these systems decrease, cost-effectiveness will improve.

A small or medium-sized utility can research state requirements and be prepared to support use of graywater by interested residences and businesses.

Rainwater

Rainwater harvesting involves the use of captured rainwater, which otherwise would have soaked into the ground, evaporated, or entered the drainage system. Rainwater is collected from roofs and flows by gravity through gutters and downspouts into a storage tank. From the tank it can be used, as is, for the landscape, or be filtered and disinfected as drinking water.

Figure 4-21. Austin City Hall in Texas collects foundation drain water and AC condensate for irrigation. (Courtesy of H.W. Hoffman, Hoffman & Associates, LLC.)

Rainwater harvesting is a form of low impact development (LID) in terms of stormwater processing. Less water goes into the municipal storm drain system because water is utilized on-site. In Green Building programs, credits can be earned by substituting rainwater for potable water for irrigation or toilet flushing. Before beginning promotion of rainwater harvesting, a utility should check state regulations.

Rain barrel is the term for the small-scale version of this technology, typically a 55- or 65-gallon plastic barrel. Cistern is the term for larger tanks, which may be inground or aboveground (Figure 4-21). Many homesteads and farms use cisterns for drinking water, and the U.S. Virgin Islands uses rainwater extensively. A 2001 book, *Rainwater Technology Handbook – Rainharvesting in Building* gives examples of rainwater use in architecturally advanced public and private buildings around the world, particularly in Germany (Konig 2001).

Many companies are now selling sophisticated aboveground plastic containers that hold 500 gallons or more. Among these are reinforced plastic bladders that can be sited under decks. An unused plastic septic tank can serve as an inground version (Hoffman 2008). A concrete tank can also be planned as part of the home's design and constructed in the ground. The entrance to an 8500 gallon lined concrete cistern that will be used for potable water was sited under the driveway in a home constructed to LEED for Homes Gold standards, shown in Figure 4-22.

One Inch of Rain on One Square Foot of Catchment Area = .6233 Gallons

The City of Austin, Texas, is fostering rainwater harvesting by providing rebates for cistern and rain barrel installation both in commercial facilities and residences. The Texas Water Development Board now has an annual rainwater harvesting competition and recognition program. *The Texas Manual on Rainwater Harvesting* is downloadable from the City of Austin website www.cityofaustin.org (Todd et al. 1997). The State of New Mexico has also published a Rainwater Harvesting Guide (New Mexico Office of the State Engineer, undated).

Figure 4-22. Belowground lined concrete cistern of 8500 gallon capacity, for indoor potable use, located under the driveway, New Smyrna Beach, Fla. Schweizer-Waldroff Architects. (Photo: Water Media Services)

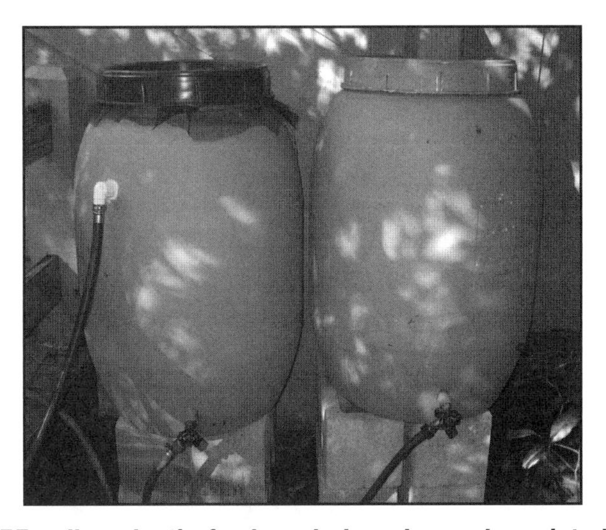

Figure 4-23. These 55-gallon plastic food grade barrels can be painted with a primer and then painted the color of the house or otherwise decorated. The screen on top prevents breeding of mosquitoes. Two or more barrels can be linked together, as shown here, with an overflow hose leading to an area of the yard needing the most water. (Photo: Water Media Services)

Some gutter pipe manufacturers include filters to facilitate rainwater harvesting. The American Rainwater Catchment Systems Association (ARCSA) (www.arcsa.org) is the most prominent organization in this field.

The larger the cistern, the greater its capacity to reduce peak summer demand. Rain barrels store so little water that they may be empty during the days of peak demand. However, they are very good as an outreach and education tool (Gauley 2008).

A small or medium-sized utility can assist interested residents to do rainwater harvesting by 1) providing information on purchase of rain barrels, 2) providing rebates or vouchers for their purchase (where an economical local vendor of rain barrels is available), or 3) offering "make a rain barrel" or "decorate your own rain barrel" workshops. Customers pay a small fee toward the barrel, spigots, glue, and screen for top. Later the rain barrels can be decorated artistically or painted the color of the house, as in Figure 4-23. Collaboration with University Extension or interested environmental groups can facilitate these workshops, and also provide an opportunity to promote other water conservation information and programs.

Water Reuse/Wastewater Recycling

The use of treated wastewater for landscape irrigation water is an increasingly common measure to reduce use of potable water. The use of *reclaimed water* is usually most feasible when a large park or golf course on the public water supply is located near a wastewater treatment plant producing a high-quality effluent. Reclaimed water traditionally is conveyed through purple pipe and should be clearly identified, as in Figure 4-24.

The lower pricing of the reclaimed water and lack of metering in early days led to high use by customers, such that back up sources of potable water needed to be made available. All reclaimed water should be metered to ensure that it is being used as efficiently as possible.

The need for reclaimed water for irrigation varies with the weather; however wastewater flows from indoor water use are relatively constant. Large storage facilities, such as reservoirs, are being constructed by some utilities to assist in capturing reclaimed water when it is not in demand.

Reclaimed water can also be used indoors for toilet flushing if it is approved by the state and if dual plumbing is installed at the time of construction. Most Green Building programs allow water efficiency credits for this indoor nonpotable use.

Car washes can use reclaimed water, and an emphasis on water reuse has occurred in car washes through sanitary sewer discharge requirements. An efficient automatic car wash often uses less water per vehicle than residents use washing a car at home. Information is available for utilities wishing to work with car washes to achieve greater efficiency (Brown 2000, Alliance for Water Efficiency 2009).

Figure 4-24. Reclaimed water signage from City of St. Petersburg, Fla., a pioneer in use of reclaimed for commercial and residential landscape irrigation. (Photo: Water Media Services)

COMMERCIAL, INSTITUTIONAL, AND INDUSTRIAL MEASURES

Commercial, institutional, and industrial users are collectively referred to as CII (commercial, institutional, industrial) or ICI (institutional, commercial, industrial). Commercial refers to all types of businesses. Institutional refers to government buildings, schools, prisons, parks, golf courses, and other public facilities. Industrial refers to industries that use water in processing.

Some of the greatest improvements in water use efficiency, along with significant energy use reduction and wastewater reduction, are possible at the CII level. CII accounts typically make up less than 5 percent of connections for a service area but use over 25 percent of the water (Hoffman 2007). These facilities often have an incentive to reduce water use because of significant water costs or waste discharge permit requirements, and providing information alone can encourage facility managers to make changes.

Training for water conservation coordinators to successfully manage CII programs is more extensive than for residential programs. The following presents strategies that may be appropriate for small and medium-sized utilities, depending on the particular service populations.

Three overall opportunities to save in this sector are 1) reducing losses (for example, performing water audits with follow-up leak repair and retrofit), 2) reducing overall water use (for example, shutting off process water when not in use and retrofitting plumbing), and 3) employing water reuse practices (for example, reusing process water) (New Mexico Office of the State Engineer 1999). For more information on CII measures, the *North Carolina Water Efficiency Manual for Commercial, Industrial, and Institutional Facilities* (1998) and the *Water Management Guide For Facility Managers of the Arizona Municipal Water Users Association* (2008) are useful references. The East Bay Municipal Utility District has teamed with CII experts around the country to publish *Watersmart Guidebook: A Water-Use Efficiency Plan Review Guide for New Businesses* (2008). This book is downloadable from the Alliance for Water Efficiency website. www.allianceforwaterefficiency.org.

As shown in Figure 4-25, the largest proportion of water use in office building, as in many other commercial facilities, is for sanitation for employees (referred to as domestic). Educating employees on water conservation is the easiest measure a commercial facility manager can perform, and utilities can provide information.

CII Audits and Retrofits

A small or medium-sized utility can follow the lead of larger utilities in the area by setting up programs to improve water efficiency equipment and practices of CII customers through audits and retrofits (Figure 4-26). If utility staff includes individuals that can receive further training and carry out the audits, then the program can be carried out in house.

Water efficiency retrofits can be managed in a similar way to which an energy service company (acronym: ESCO or ESCo) works to improve energy efficiency. A third-party contractor designs, installs, finances, and, if required, operates a new technology and is paid according to the savings achieved (i.e. the performance). These water efficiency contractors targeting CII customers are called WASCOs (water service

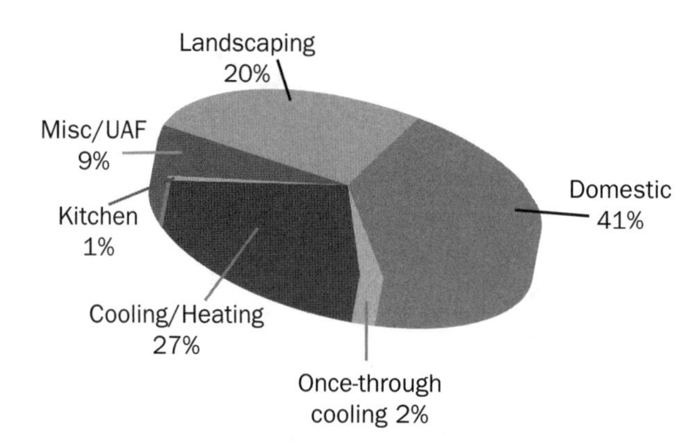

Figure 4-25. Water Use Distribution of a typical office building. Domestic (toilets, urinals, faucets, etc.), Cooling/Heating, and Landscaping are the three uses offering the best opportunities to conserve. (Federal Energy Management Program 2009)

companies). In performance contracting, the contractor is paid for its services through savings the institution will obtain following retrofits recommended by the contractor's audit. The fact that companies would risk being paid under this arrangement shows the reliability of the water savings.

Identifying locally available WASCOs and recommending these to CII customers is a viable method to achieve savings for a small or medium-sized utility with high water use by CII customers.

High-Efficiency Urinals and Flushometer Toilets

The USEPA estimates that of the 12 million urinals in the U.S., up to 80 percent are inefficient units that use as much as 4 gpf. The Energy Policy Act of 1992 established the maximum flush volume for all urinals manufactured in the United States after January 1, 1994, at 1 gpf.

Urinal models that use 0.5-gallon per flush or less have done well in performance tests, and a new category of urinals has been defined, the High-Efficiency Urinal or HEU (USEPA 2009), which include nonwater-using urinals as well as urinals using as little as one pint (.125-gallon) per flush. Utilities should focus on recommending or rebating urinals that are third-party performance tested and labeled by WaterSense, under a specification to be available in 2010.

For Flushometer toilets commonly used in CII applications, a 1.28-gpf effective flush volume should become the new standard (Alliance for Water Efficiency 2009). Testing for actual flush volume as part of a CII audit, conducted by Water Management Inc, one of the oldest WASCOs, is shown in Figure 4-26.

A manual dual-flush handle installed on new toilets or retrofitted onto 1.6 gpf toilets allows the user to choose a high or low volume flush and has been promoted for Green Buildings. Installed in the Portland, Oregon airport, it is part of a "conservation ethic," demonstrating stewardship of resources, as shown in Figure 4-27. However, currently there is no evidence that the dual-flush handle retrofit actually reduces water use (Koeller 2009).

Figure 4-26. Flushometer toilet flow rate being tested as part of CII audit. (Photos: Water Management Inc.)

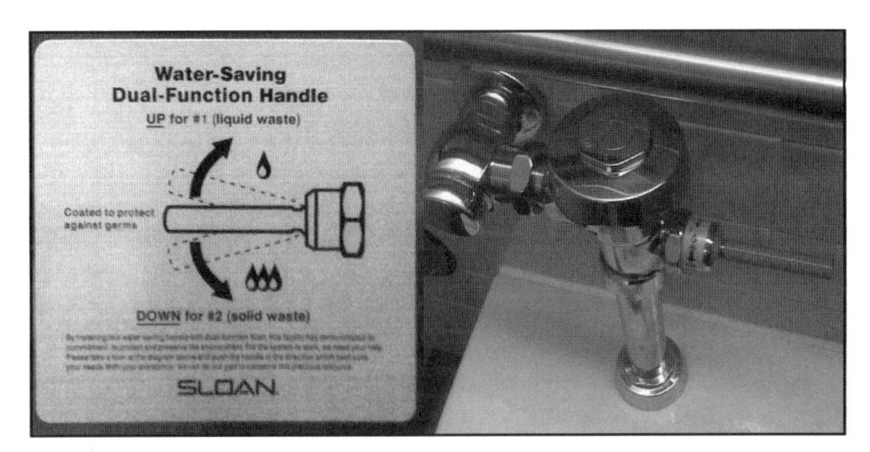

Figure 4-27. Flushometer toilet flow with Sloan Upper Cut handle that converts 1.6 gpf flushometer to dual-flush, Portland Airport, Ore. (Photos: Water Media Services.)

Many infrared sensor-operated toilets and faucets do *not* reduce water use, as documented by several recent studies. These automatic devices are correctly promoted for their health value, in that touching the handle is eliminated, but they are not correctly promoted for their water conservation value (Koeller 2007). Newer models may be more efficient, but a utility is cautioned to avoid use of and recommendation of sensor-operated products until they have been tested and certified through WaterSense.

A small or medium-sized utility can work with designers of new commercial buildings to help them find accurate product information as they follow the water efficiency portion of the Green Building standards. The utility can also inform building managers how to find new water efficiency research. The Alliance for Water Efficiency website is the source for up to date information on product studies (www.allianceforwaterefficiency.org), along with the website of the California Urban Water Conservation Council (www.cuwcc.org) and USEPA WaterSense program, www.epa.gov/watersense.

Commercial Faucets

Commercial faucets are specified to use no more than 0.5 gpm, following an ASME standard adopted by the model plumbing codes by reference. This low requirement would indicate that faucet retrofit efforts by the utility in the commercial sector would not be necessary; however, many design engineers, specifiers, plumbing contractors, and building owners have incorrectly interpreted codes and installed 2.2 gpm faucets in commercial facilities (Koeller 2007).

The 2.2 gpm standard applies only to faucets used in "private" settings, including residences, hotel/motel guest rooms, and private rooms in hospitals, as defined by the Uniform Plumbing Code, International Plumbing Code, and the National Standard Plumbing Code. All other lavatories are public and must follow the 0.5 gpm standard.

For most incorrectly installed commercial faucets with flow rates above 0.5 gpm, installation of low flow aerators is an easy solution. Unlike low flow showerheads, which met resistance in the early years, low flow faucets and faucet aerators have generally been very well received. The alternative to a faucet with aerator is splashing, which is unpopular. Because faucets are often used for only 5 to 30 seconds, water savings through retrofits are small when compared to replacing toilets. But replacing aerators is cost-effective as it generally costs less than $1 per faucet (Alliance for Water Efficiency 2009).

Sensor-activated faucets do not save water according to studies compiled by the California Urban Water Conservation Council and Alliance for Water Efficiency (Koeller 2007, Gauley 2008). Sensor-activated faucets can be correctly promoted for their health benefits in being "touchless" but are not correctly marketed for water conservation benefits.

Many hospitals restrict the use of aerators and mandate that only laminar flow controls be used. Laminar flow controls prevent airborne spread of water droplets to minimize a patient's exposure to bacteria such as Legionella. The suggestion is to use laminar-flow faucets that use no more than 1.5 gpm, where required in medical facilities (Water Management Inc 2009, EBMUD 2008).

Small and medium-sized utilities can provide information to commercial facilities and possibly incentives for retrofits.

Pre-Rinse Spray Valves

Pre-rinse spray valves, as shown in Figure 4-28, are used in almost all restaurants, including fast-food restaurants and grocery store delicatessens, to rinse pans and dishes prior to loading into automatic dishwashers. They are also used in cafeterias of schools, prisons, and similar institutions. Pre-rinse spray valves manufactured on or after January 1, 2006 must have a flow rate of no more than 1.6 gpm; however many older models use over 5 gpm. New requirements are to clean food off a plate in 26 seconds.

Improved efficiency of the new spray valves allows workers to spend less time preparing dishes for the dishwasher. Plus there is significant savings in energy for heating the water. Consequently, restaurant managers have received retrofits very well.

Figure 4-28. Pre-Rinse Spray Valves. A maximum flow rate of 1.6 gpm has been required since 2005 under EPACT amendments. Even lower volume has also been shown to function well. (Photo: Maddaus Water Management)

A research center funded by Pacific Gas & Electric (PG&E) called the Food Service Technology Center in San Ramon, California, identified pre-rinse spray valves as one of the easiest to implement and most cost-effective CII water efficiency opportunities. Methods for pre-rinse spray valves distribution include rebates or vouchers, dealer/distributor incentive programs, and installation as part of a commercial audit program.

The California Urban Water Conservation Council has determined that door-to-door direct installation of a free pre-rinse spray valves is the most effective means of implementation. Cost of the devices is under $60 each (which is no more than higher flow fixtures commonly sold). Life of the measure is considered 5 years (Chesnutt et al. 2007). With cooperative funding from energy utilities, many water utilities are now providing free pre-rinse spray valves to restaurants and institutions.

In 2009, USEPA WaterSense and ENERGY STAR Programs agreed to collaborate with ASME and ASTM to develop performance and efficiency criteria toward an USEPA specification for pre-rinse spray valves.

A small or medium-sized utility in areas with many restaurants should investigate a pre-rinse spray valve program. The utility can possibly join with a larger utility that may be administering the program, seek joint funding with an energy utility, and/or contract with one of the water conservation vendors to administer the program.

Commercial dishwashers also have potential for water and energy efficiency improvements that can be identified in a CII audit.

Air-cooled Ice Machines

Water-cooled ice machines have been preferred by hotels and restaurants, as they were thought to use less energy than air-cooled. However, when the cost of energy used to produce the water (called the *embedded energy*) is included, air-cooled machines clearly use less energy and are the better choice (Hoffman 2007). A small or medium-sized utility in a high tourism area can provide this information to hotels and restaurants.

Commercial Washing Machines

Clothes washers located in a common area of an apartment building provide excellent opportunities for efficiency improvements. These may serve multiple apartments and are usually coin-operated appliances. As for residential washers, these commercial clothes washers have improved in water efficiency, thanks to efforts of energy and water utilities working with the Consortium for Energy Efficiency (CEE). Seattle Public Utilities and other larger utilities give rebates for retrofit to more efficient washers, listed on the website www.cee1.org. Seattle Public Utilities encourages users of coin-operated washers to load machines to full capacity rather than doing a series of smaller loads

Machines used in on-premise laundries, including those of large hotels, and industrial laundries are entirely different pieces of equipment. The Federal Energy Management Program, Water Efficiency Section, of the US Department of Energy has compiled information on improving efficiency of the latter types of machines. Water recycling and ozone disinfection are two water efficiency measures. The FEMP website is www1.eere.energy.gov/femp.

A small or medium-sized utility that has laundromats and commercial laundries among its high water users can work with the owner to identify savings opportunities.

Cooling Towers

Most high-rise office buildings, condominiums, and hotels have cooling towers, as shown in Figure 4-29. These account for a significant percentage of the building's total water use.

To remove the same heat load, single-pass systems use 40 times more water than a cooling tower operated at five cycles of concentration. *Cycle of concentration* is a measure of cooling-tower operation efficiency comparing water going in and out. The State of Arizona requires cooling towers to use at least three cycles, and six or more cycles are recommended by the industry. Innovative technologies currently allow up to 50 cycles of concentration, depending on the type of cooling tower and the facility management (Puckorius 2008).

In simplified terms, warm water from air conditioning, refrigeration, or industrial processes enters the cooling tower system and is allowed to evaporate, leaving the system water more concentrated with dissolved minerals. To prevent suspended and dissolved solids from building up and causing scaling and corrosion, additional water must be added (make-up), as in Figure 4-30, and some of the water in the basin periodically discharged (blowdown or bleed-off). Releasing and replacing small amounts of water continuously, rather than releasing large amounts of water periodically can reduce bleed-off. Adding special chemicals to the water that allow higher concentrations of dissolved solids is an additional way to reduce bleed-off. Metering the quantity of water put into and discharged from the cooling tower allows better management of the efficiency of the tower (Vickers 2001, US EPA 2009). Alternate nonpotable water sources can also be used for make-up water.

Cycles of concentration are measured in terms of total dissolved solids (TDS). Electrical conductivity, which is related to TDS, in microsiemens, is often used in place of TDS. For example, if the conductivity of the make-up water were 100 microsiemens

Figure 4-29. Cooling towers, such as this one, consume 20 to 30 percent or more of a facility's total water use. (Photo: Water Management Inc).

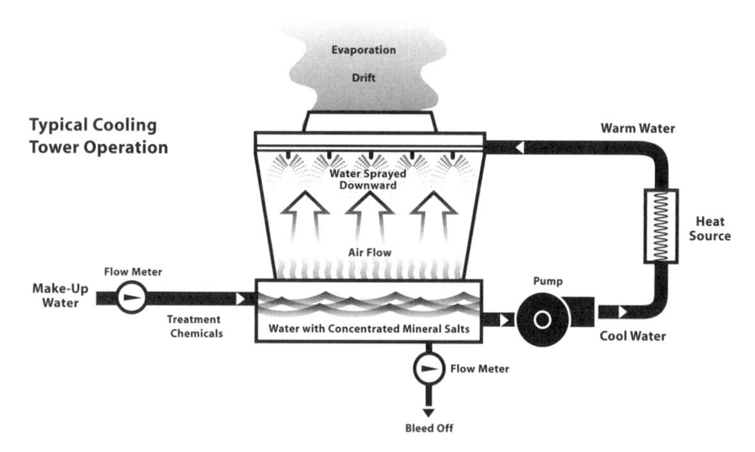

Figure 4-30. Schematic of cooling tower (Diagram Paul Puckorius, Graphic: Water Media Services)

while that of the bleed-off water is 500 microsiemens, the tower would be operating at five cycles of concentration. The quantity of bleed-off can be controlled using an automated conductivity controller, to monitor optimal cooling tower water chemistry (EBMUD 2008, 2009).

Several large utilities provide generous rebates for conductivity controllers and flow meters installed to monitor cooling tower make-up and bleed-off. Among these are the Metropolitan Water District of Southern California through its Save Water, Save A Buck Program for Southern California businesses, Seattle Public Utilities with its Saving Water Partnership, and Denver Water.

Small and medium-sized utilities in areas with large numbers of cooling towers can profit by working on a cooling tower program, particularly if air conditioning in large buildings in the utility service area contributes to peak demand. The property owner will achieve water and energy savings through cooling tower efficiency improvements, so it may not be necessary to provide financial assistance, only education.

Small and medium-sized utilities might work with larger utilities to organize cooling tower workshops, taught by experts in the field and targeting property owners and managers. The AWWA Section water conservation committee and the Alliance for Water Efficiency can be consulted for local availability of these workshops.

COMMERCIAL LANDSCAPE MEASURES

Commercial and institutional facilities may have large landscaped areas utilizing in ground automatic irrigation systems. Many of the techniques for reducing irrigation water use for residential customers apply here, but large-scale landscape water users may also benefit from central control of sprinkler systems. More easily than residential users, they may be able to afford the costs of highly efficient irrigation systems, including careful design, good maintenance and *SMART controllers*.

Large Landscape Irrigation Audits

For small commercial sites, the same or similar lawn-watering guides can be provided as for residential. The landscape manager or property owner can follow the instructions to perform a simple water audit by measuring the distribution uniformity and precipitation rate of the existing sprinkler system and adjusting watering times.

For large landscapes, trained auditors examine the irrigation system, note any repairs and/or changes that should be implemented, and prepare a customized irrigation schedule for the site, as shown in Figure 4-31. The objective of the program is to provide landscape managers with the information they need to perform timely equipment maintenance and apply accurate amounts of water throughout the irrigation season.

The auditor should be an Irrigation Association Certified Landscape Auditor. A background in the irrigation industry is beneficial, because a wide variety of irrigation equipment will need to be evaluated by the auditor.

The utility should

- Identify and prioritize sites for auditing, based on irrigated acreage and past water use.
- Mail informational letters about the audit program and commercial irrigation guides to potential sites.
- Answer customer questions by phone or email, and set up the appointment for the audit.
- Perform audits of selected sites. Auditors check for leaks, mismatched sprinkler heads, and overspray (watering streets and sidewalks, overlapping sprinkler areas).
- A customized irrigation schedule with recommendations for repairs and/or retrofits is provided to the building owner or landscape manager.
- Continue follow-up support for the program by providing weather information for updated irrigation schedules and other topical information.
- Tie into a local weather station offering evapotranspiration (ET) data and arrange to make this information available to irrigators.
- Provide ongoing follow-up support for the customer by periodically providing data on savings.

Following the recommendations of landscape audits can result in an estimated 15 percent water use reduction. The total cost of the delivered audit will vary, depending on the size of the lot and the complexity of the existing irrigation system. Audit costs are lower for large, uniform areas (such as playing fields) and higher for complex parceled sites (such as condominium developments). Ongoing costs for the program involve follow-up calls or visits by the auditor, which are recommended.

To find auditors in your area, visit the WaterSense site or the Irrigation Association website, www.irrigation.org. Larger utilities in a specific area should be consulted about typical costs.

SMART Controllers

For more than 20 years, golf courses, parks, and athletic departments have used state of the art irrigation technologies that automatically adjust watering schedules based on local weather and site conditions. Unlike irrigation timers that water on a set schedule, these so-called SMART controllers use weather, site, or soil moisture data to determine an appropriate watering schedule. These sophisticated add-ons to the irrigation system are of two types: 1) climate-based, "ET- or Evapotranspiration-based, and 2) soil moisture sensor-based controllers.

Evapotranspiration (ET), is defined as the quantity of moisture that is both transpired by the plant and evaporated from the soil and plant surfaces. ET is a function of weather conditions and plant type. Most of the climate-based products use real time or historic weather data to schedule irrigation. ET-based controllers use data from local weather stations, gather information from several weather stations linked to match the installation site's zip code and address, or use on-site sensors.

Several irrigation equipment manufacturers use the WeatherTRAK signal and algorithm that takes weather station data and runs through proprietary software. Other manufacturers sell systems with weather sensors that gather onsite weather data, as in Figure 4-32. These weather stations measure solar radiation, wind speed, temperature, and humidity. Solar radiation is considered the most important factor in ET, followed by wind.

A 4-year research study on Smart Controllers by the Metropolitan Water District of Southern California with 26 Southern California water providers and the East Bay Municipal Utility District with 5 Northern California water providers, concluded in 2009. The study found that most SMART controllers brands and technologies reduced demands on average, but not all reductions were statistically significant. At sites that had historically under-irrigated, the SMART controllers actually increased water use. Conclusions were that water savings using these controllers can be maximized by improved programming and by targeting over-irrigators (Mayer et al. 2009).

Soil moisture sensor (SMS) controllers measure soil moisture content in the active root zone. If the SMS controller determines that soil moisture content is above the user-adjustable soil moisture content set point, the preset irrigation cycle will be bypassed. In this way, the SMS functions similarly to a rain sensor that bypasses irrigation during and following rainy periods. Beyond what the rain sensor does, however, the SMS can detect water from irrigation, and when enough has been applied, either interrupt the cycle or adjust the run time to maintain the desired level of soil moisture (Philpott et al 2008). Contractors need to be trained in correct installation, however.

Figure 4-31. Commercial irrigation audit evaluates distribution uniformity and other measures of efficiency. Audit is appropriate prior to installation of SMART controllers or other system enhancements. (Photo: Water Management Inc.)

Benefits of both ET controllers and SMS controllers are more efficient water use, reduced runoff, better plant health, and customer convenience. Although SMART Controllers may be too expensive for widespread retrofits in the residential market, they are appropriate for commercial projects or subdivision common areas. In the residential sector, they could be promoted to the top tier of water users who contribute to high system peak season water use. Several large utilities are giving rebates for installation of SMART Controllers (Aqua Conserve 2002, Ash 2002).

It should be noted that models that work well in one part of the US might not in another, due to soil type and weather cycle differences. This is why the SMART Water Application Technology (SWAT) program's work testing these products, and WaterSense's later certifying these controllers is so important. For highest degree of success in promotion of SMART controllers, disparate stakeholders need to be educated, high water users need to be targeted, and there must be high operator integrity (Chesnutt et al. 2007, Mayer et al. 2009).

A small or medium-sized utility can follow the lead of larger utilities in the area regarding promotion of SMART Controllers.

INDUSTRIAL CONSERVATION MEASURES

Water is used by industries for employee sanitation, cooling, fume and gas scrubbing, process uses, and landscaping. The production of some goods (such as food and beverages, paper products, and microchips) is highly water intensive. Opportunities for water conservation are generally site-specific.

Optimizing flow rates and maximizing the recycling of process water can achieve conservation. For example, water used for cooling that is not significantly degraded in quality may subsequently be used for irrigation, flushing toilets, or processes that do not require potable water.

Figure 4-32. ET controller with on-site weather that uses solar radiation, temperature, and rainfall to modify irrigation system scheduling. Controller is Solar Sync by Hunter Industries, photographed at SeaWorld, Orlando, Fla. (Photo: Brian Walker)

Once-through (single-pass) cooling with potable water is where water is circulated once through a piece of equipment and is disposed of down the drain. The types of equipment that typically use single-pass cooling include: CAT scanners, degreasers, hydraulic equipment, condensers, air compressors, welding machines, vacuum pumps, ice machines, x-ray equipment, and air conditioners (EBMUD 2008, 2009). To maximize water savings, single-pass cooling equipment should be either modified to recirculate water or if possible, should be eliminated altogether.

Most industries have significant incentive to conserve water. The economic benefits of conservation include reduced water charges, lower costs for pretreatment and wastewater treatment because of reduced wastewater flows, and reduced energy costs for hot water. A small or medium-sized utility might provide information to local industries identifying opportunities or encourage use of performance contractors (WASCOs).

Potential Water Savings

5

SUMMARY:

- *Describes an example conservation program for small utilities*
- *Includes checklists that show which conservation measures are appropriate for minimum, moderate, or maximum conservation programs*

The following summarizes the potential for utility (system) water conservation measures (Chapter 3) and measures for residential, commercial, institutional, and industrial conservation customers (Chapter 4). For consumer measures, the small or medium-sized utility will be seeking primarily voluntary participation, because paying to help customers replace fixtures through rebates may not be cost-effective. Voluntary customer participation will be increased if the utility has volumetric pricing and steeply tiered rates. Conservation potential is high where the combined water and wastewater costs are high, for example greater than $3 per 1,000 gallons, at use volumes higher than 5000 gallons.

Previously, the amount of water savings achievable through specific residential conservation measures was very difficult to verify. Currently, however, utilities can refer to benchmarking data generated in the AwwaRF's 1999 *Residential End Uses of Water* and later studies. The California Urban Water Conservation Council and the Alliance for Water Efficiency have calculation tools available referencing the latest water savings information. Some state agencies may have tools as well.

POTENTIAL FOR UTILITY (SYSTEM) CONSERVATION MEASURES

If nonrevenue water is above 10 percent of total production, it should be possible to reduce it significantly. Metering previously unmetered customers can reduce consumption by 20 percent (along with appropriate pricing). If nonrevenue water is low, and metering is complete, the potential for utility (system) level savings will be low (around 5 percent).

Implementing water conservation rates may reduce water use by 5–10 percent. The potential to continue reductions through raising the higher rate tiers continues in most cases. A utility's public information programs can reduce consumption up to 5 percent. Together, these utility measures can be very effective, but the these savings are not additive and depend on the specific situation.

POTENTIAL FOR RESIDENTIAL CONSERVATION MEASURES

A 15 percent reduction in residential water use over a 10-year period is a reasonable minimal goal for a first-time conservation program. Such a goal would suggest that a conservation program include requirements for fixtures that exceed EPAct 1992 specifications, an aggressive fixture retrofit program (for which funds are budgeted), and a landscape water conservation program.

Higher reduction goals are appropriate for currently unmetered communities (with metering and volumetric pricing to be put into place), or where water shortages are acute and the avoided cost of water is high. Lower goals are reasonable for communities where extensive conservation is already in place, where water shortages are not severe, or where water is inexpensively priced.

POTENTIAL FOR COMMERICAL, INSTITUTIONAL, AND INDUSTRIAL CONSERVATION MEASURES

Each community's CII customer base is different, so estimates of potential use reduction vary widely. Toilets, urinals, faucets and landscaping all provide opportunities for water savings for CII customers. Weather-based SMART controllers and central control systems that are beyond the price range of the average homeowner may be cost-effective for large landscapes and may allow significant use reductions. Combined water, wastewater, and energy costs can be significant motivating factors for these entities to make process changes.

Determine the Conservation Potential of Specific Measures for a Service Area

Key pieces of information needed to generate in Chapter 2 are who are the high water use customers, what age is the housing stock, and what types of businesses, industries, and institutional buildings are in the service area. Peak use patterns have also been generated.

A list of most promising customer measures can be made. As a later step, a utility can prioritize these most promising measures and bundle them together into alternative water conservation plans with attached proposed budgets. Estimating potential water savings, the measures that will give the most return from the investment can be determined.

For most calculations, AWWA M52 *Water Conservation Programs—A Planning Manual* was used as a reference, but one basic formula will determine how effective a specific conservation measure might be in a given year, for a particular group of water users.

Water savings resulting from conservation measures will depend on (Maddaus 2006)

(1) The reduction in water use as a result of implementing the measure, and
(2) The degree of coverage that the measure can achieve (also known as *market penetration*).

$$E = R \times C \times Q$$

where: E = reduction in water use as a result of the measure, in millions of gallons per year (mgy), for the year of interest.

R = reduction in water use as a result of the measure, expressed as a fraction of 1.

C = percent coverage of the measure for the group of water users under consideration (market penetration), for the year of interest. Also called the installation rate.

Q = baseline water use for the group of interest, before conservation measure was put into place (mgy), for the year of interest (from Worksheet 1).

Determine R, the fraction reduction in water use.

The factor R is the fractional reduction in water use that is expected to result from a particular conservation measure. To estimate this reduction, you need information on the actual water savings as well as information about the average water use for the user group in question. The fractional water savings can then be estimated by the formula:

$$R = S/W$$

where: R = fraction water reduction for the year of interest (to calculate)

S = water savings resulting from the measure, in gpd, from Table 4-1, Chapter 4 and up-to-date reports from the Alliance for Water Efficiency

W = average water use without the conservation measure in place (gpd), for the year of interest (from Worksheet 1)

Two examples are provided.

Worksheet 3
Example 1
Evaluate Conservation Potential (Use Savings Rates from Table 4-1)

Measure Description: Toilet replacement in multi-family housing complex with pre-1985 toilets (using 5-7 gpd av. 6 gpd). HETs (1.28 gpf) will be installed. Average occupancy is 4. Number of units is 200.

Fraction reduction in water use, *R*:	
where *R* = *S*/*W*	
Average use with conservation, *S*	1.28 gpf × 4 persons × 5.1 flushes = 26.1 gallons per household day
Average use without conservation, *W* (gpd)	6 gpf × 4 persons × 5.1 flushes= 122.4 gpd
Fraction reduction in water use, *R*	21% water use reduction
Market penetration/installation rate	C: 100% installation rate
Baseline water use, *Q*:	3 kgal/mo × 12 mo × 200 units =7.2 mgy
Overall reduction in water use, *E*:	
where *E* = *R* × *C* × *Q*	21% × 100% × 7.2 mgy =1.4 mgy
Percent reduction in water use:	
where % reduction = (*E*/*Q*) × 100	21% reduction in target 200 units

Worksheet 3
Example 2
Evaluate Conservation Potential (Use Savings Rates from Table 4-1)

Measure Description: Rain sensor installation for the 5,000 homes in service area with automatic irrigation systems, through ordinance change and enforcement. To educate, create awareness, and assist customers who would have difficulty complying, 300 rain sensors will be given away at a cost of $4,500. This analysis will not look at cost-benefit but estimated water use reduction from the measure, including ordinance, enforcement and giveaway (Reduction estimate from Table 4-1*)

Fraction reduction in water use, *R*:	
where *R* = *S*/*W*	
Average use with conservation, *S* (gpd)	______________
Average use without conservation, *W* (gpd)	______________
Fraction reduction in water use, *R*	20% water use reduction estimated *
Fraction coverage of measure (market penetration/installation rate), *C*:	60% market penetration estimated
Baseline water use, *Q*:	6,000 gal/mo × 12 mo × 5000 = 360 mgy
Overall reduction in water use, *E*:	
where *E* = *R* × *C* × *Q*	20% × 60% × 360 mgy = 43.2 mgy
Percent reduction in water use:	
where % reduction = (*E*/*Q*) × 100	12 % overall reduction in target 5000 homes

(continued next page)

WORKSHEET 3 (continued)
Evaluate Conservation Potential (Use Savings Rates from Table 4-1)

Measure Description: ______________________________	
Fraction reduction in water use, R:	
where $R = S/W$	
Average use with conservation, S (gpd)	____________
Average use without conservation, W (gpd)	____________
Fraction reduction in water use, R	____________
Fraction coverage of measure (market penetration/installation rate), C:	____________
Baseline water use, Q:	____________ mgy
Overall reduction in water use, E:	
where $E = R \times C \times Q$	____________ mgy
Percent reduction in water use:	
where % reduction $= (E/Q) \times 100$	____________ %

EXAMPLE CONSERVATION PROGRAM FOR A SMALL UTILITY

This example program has five elements with measures typically applicable to the utility and to specific user categories: single-family residences, multifamily residences, commercial and institutional facilities, and industrial facilities. This would be considered a minimum program in terms of funding with a budget of $1,000-$10,000. The program carefully targets the high water users and most applicable sectors. The following are measures carried out by this example small utility.

Element 1: Utility

- Meter all customers.
- Utilize volumetric pricing (tiered water rates), with informative billing and periodic rate increases for the higher tiers.
- Reduce nonrevenue water through audit, metering, leak detection and repairs.

Water savings of this element could be as high as 5–10 percent of total production, with percentage reduction increasing with further expenditures (for metering and leak repair).

Element 2: Single-Family Residences

1. Provide public information on water supply issues, and the utility's conservation program annually in print as the Consumer Confidence Report and on the utility website.
2. Collaborate with other nearby utilities in public information campaigns, focused on irrigation reduction during the dry season and in winter, if in a Sunbelt state where winter watering is common.

3. Perform school visits on career day and collaborate with other utilities to offer teacher training using one of the commonly available in-school education programs.
4. Print and distribute a residential lawn-watering guide, and distribute information in city newsletters and on websites.
5. Become a USEPA WaterSense promotional partner, at no cost, and publicize water savings that customers can achieve through new fixture installation.
6. Provide low-flow faucet aerators and retrofit kits at convenient depots, along with water conservation print literature, referencing WaterSense.
7. Perform free indoor water audits. The audits will identify indoor water use habits, repair leaks, and install low water-use hardware (showerheads, aerators, toilet devices).
8. If landscape water use with automatic irrigation system is a major use sector, seek funding to provide free or low cost landscape audits. The audits will propose landscaping changes such as the use of low water-use landscaping techniques and plants and the use of efficient irrigation systems and practices.
9. Modify local ordinance language for new construction or (as an interim step) provide builder incentives for such measures, for example fast-track permitting, lower impact or hookup fees.
   - Require installation of HETs (1.28 gpf) and other WaterSense labeled products through a local or state regulation.
   - Prohibit multiple showerheads in new construction.

The major cost will be for staff time to coordinate programs, including performing residential water audits and follow-up tasks (leak repair and fixture installation) and for purchase of retrofit items or kits. The same items or kits can be used in the multi-family sector. A minor amount of staff training will be necessary to perform indoor water audits (suggested audit procedure appears in Appendix B).

Other costs are contribution to printing lawn-watering guide, shared with other utilities. For a new program, demand can be reduced at least 10 percent of preconservation levels.

Element 3: Multifamily Residences

The example utility has a federally funded low income housing complex that is master metered. Performing a water audit for this and other multifamily dwellings, if followed by retrofits, reduces many users at one time. The utility will

- Provide public information on water supply issues and the utility's conservation program to the building owner, manager, and residents.
- Perform free indoor water audits for older multifamily units. Identify indoor water use habits, repair leaks, and install low water-use hardware (showerheads and aerators). The auditor will collect data on toilet flow rates and frequency of leaks prior to seeking funding for a toilet retrofit program for the building.
- Provide irrigation schedules for landscaped areas, and propose the use of low water-use landscaping and efficient irrigation systems and practices.
- Perform free landscape water audits for multifamily units with significant amounts of landscaping. Audit staff can meet with building owners and managers to provide information on savings that could be achieved through irrigation equipment replacement and jointly seek funding for a retrofit program.

- Require use of HET (1.28 gpf) and low-flow showerheads (below 2.5 gpm) for any new construction, through local or state regulation. Or provide builder incentives for such measures, for example fast-track permitting, lower impact or hookup fees.

For a newly started program, this program can reduce demand at least 10 percent of pre-conservation levels.

Element 4: Commercial and Institutional Facilities

- Identify high water use commercial and institutional customers and note if use correlates with peak demand. Taper program to include irrigation programs accordingly.
- Provide public information on water supply issues and the utility's conservation program to the building owner, manager, and employees.
- Provide free water-use audits for larger commercial and public facilities. Identify and repair plumbing leaks and install water-saving fixtures for bathroom use. Focus on the fixtures audited in multi-family also.
- Have audit staff meet with building owners and managers to provide information on savings that could be achieved through fixture replacement and jointly seek funding for a retrofit program.
- Encourage irrigation scheduling for smaller sites by providing a lawn-watering guide for turf areas and providing information on how to measure precipitation rates from sprinkler systems.
- Encourage building owners or managers to consider contracting with a performance contracting company (WASCO) to reduce water use.
- For restaurants, school and prison cafeterias, encourage the use of, and if possible, provide low-flow prerinse spray valves. Seek cost-sharing with local energy utilities.
- Work with state agencies and other utilities to set up a cooling tower workshop.

Water savings for this element will be moderate to low, less than 10 percent of pre-conservation levels. If the entities contract with a WASCO and implement equipment and process changes, savings will be greater.

Element 5: Industrial Facilities

An industrial conservation program requires specialized knowledge about industrial processes. The degree to which a utility should focus on this sector is determined by whether the customer is one of the utility's highest users and if use is correlated with peak demand. Water rates should be structured to provide these large water use customers a strong incentive for conservation.

The utility can encourage facility managers to consider contracting with a performance contracting company to reduce water use. It will seek opportunities for use of reclaimed water or other nonpotable water to substitute for potable water for processes.

Summary

This example program will not require major expenditures by the utility, particularly if knowledgeable in-house staff can carry out audits. The primary disadvantage of the program is that the water savings will be modest, less than 10 percent of preconservation levels. This reduction may be enough, however, depending on the utility's conservation goals. A more aggressive program can focus on the major water user categories within a service area and seek partnership mechanisms to achieve retrofits.

Table 5-1 summarizes the measures previously discussed, with emphasis on what can be accomplished by small and medium-sized utilities. Checklists 1 and 2 divide potential measures into minimum, moderate and maximum, as discussed previously (minimum $1–10,000 project budget, moderate $10,000–100,000 project budget, and maximum project budget above $100,000). This example program falls between the minimum and moderate program shown in the checklists.

Table 5-1. Example Water Conservation Program for a Small to Medium-sized Utility

SECTOR	PROGRAM
Utility	Metering System water audit and reduction of nonrevenue water Conservation pricing Public information, including in-school education Setting good examples for the public in all utility buildings and municipal landscapes
Single-Family Residential	High Efficiency Toilets (customer information only, no rebates for a small utility) Low-flow showerhead giveaway/exchange Retrofit kits Water audits (medium-sized or large only) Lawn-watering guides
Multifamily Residential	High Efficiency Toilets (information only for a small utility) Low-flow showerhead giveaway (manager installation) Water audits (medium-sized or large only) Retrofit kits
Commercial and Institutional Facilities	Employee education Water audits, including large landscape Lawn-watering guides
Industrial Facilities	Lawn-watering guides Employee education Water audits, including cooling towers (large only)

Not all of these measures can be accomplished with staff and financial limitations. Measures included on this list but not checked are planned if increased funding becomes available.

CHECKLIST 1
POTENTIAL UTILITY CONSERVATION MEASURES (Chapter 3)

The checklist summarizes potential conservation measures that can be carried out by the utility, and shows whether they are appropriate for minimum, moderate, or maximum conservation programs. (Minimum programs would typically have a $1-10,000 project budget, moderate $10,000 to $100,000 project budget, and maximum project budget over $100,000).

	Consider For:		
Potential Measure	**Minimum Program**	**Moderate Program**	**Maximum Program**
System Leak Reduction Achieves year-round conservation Requires utility water audit	X	X	X
Metering Potentially high water savings New construction less expensive than retrofit	X	X	X
Pricing Increasing block rate or seasonal changes Requires meters Requires approval by regulatory body	X	X	X
Public Information For both peak-use and year-round Most useful in conjunction with other measures	X	X	X
Incentives and Regulations Incentives include fee discounts and customer rebates Regulations can mandate types of device or amount of water used		X	X

CHECKLIST 2
POTENTIAL CONSUMER CONSERVATION MEASURES (Chapter 4)

This checklist summarizes potential consumer conservation measures and shows whether they are appropriate for minimum, moderate, or maximum conservation programs.

Potential Measure	Minimum Program Budget $1,000–$10,000	Moderate Program Budget $10,000–$100,000	Maximum Program Budget over $100,000
Indoor Residential			
Giveaway of retrofit kits	X	X	X
Faucet aerator giveaway	X	X	X
Indoor water audit and leak repair		X	X
Low-flow showerhead, giveaway/exchange		X	X
High efficiency toilets rebates		X	X
High efficiency clothes washer rebates			X
Outdoor Residential			
Lawn-watering guides	X	X	X
Promotion of low water-use landscaping		X	X
Landscape irrigation ordinance			X
Commercial and Institutional			
Large landscape irrigation audits			X
Landscape ordinance			X
Provision of reclaimed water			X
Prerinse spray valve replacement		X	X
Industrial			
Reclaimed water			X
Landscape irrigation reduction		X	X
Improved plant facilities maintenance			X
Cooling tower improvements			X

Designing a Conservation Program

6

SUMMARY:

- *Provides guidelines for designing a conservation program that can meet a utility's specific needs and goals.*

The following activities have been completed:

- Reviewed the reasons why conservation may be an appropriate tool for a utility (Chapter 1)
- Identified a utility's water use characteristics related to potential water conservation measures (Chapter 2)
- Reviewed potential utility (system) conservation measures (Chapter 3)
- Reviewed potential consumer conservation measures, including residential and CII (Chapter 4)
- Considered potential water savings that can be realized from these measures and reviewed an example conservation program (Chapter 5)

This chapter provides guidelines for designing a conservation program that can meet a utility's specific needs and goals. Appendix D presents an outline for a typical water conservation plan (Maddaus 2006).

PROVIDE FOR CITIZEN INVOLVEMENT

Conservation can save utilities money by delaying the construction of capital facilities, reducing energy consumption, and reducing wastewater flows. Conservation can also help extend inadequate water supplies and alleviate future water demand caused by population growth. Consumers can benefit by saving money on water and energy bills.

Although a utility may aware of conservation savings, the public may not. An important early step in designing a conservation program is to enlist the involvement and support of the community. Conservation must have widespread support to be fully effective.

In small communities, there may be already close communication between the utility and elected officials. This communication plus the formation of an informal advisory panel may be sufficient. In larger urban areas where close informal contact is more difficult to maintain, an official committee should be formed to advise the utility on the conservation program and to provide feedback.

A conservation network should include (Maddaus 1987a)

- Elected officials from jurisdictions affected by conservation
- Staff from private water companies
- Staff from local governmental agencies
- Staff from state agencies
- Representatives from major interest groups likely to be affected by conservation: industry, the chamber of commerce, builders' associations, farm bureaus, fisheries cooperatives, tourism boards, realtor boards, and landscape contractors
- Community representatives: civic associations, neighborhood associations, local press, and media owners
- Local environmental interest groups
- Local professionals with some technical credibility: economists, engineers, and planners
- Representatives of major water users, such as food processing plants and homeowners associations.

Public meetings and publicity should be scheduled at key points in the conservation planning process to allow input on the decisions being made. Staff time should be designated to set up and conduct these meetings and to prepare publicity about the program.

As important as it is to educate the public, it is just as important to listen to the public. A program that does not satisfy public needs will be difficult to maintain. A customer survey may be a productive way to solicit input from the public on conservation goals and attitudes (Example in Appendix C).

RESEARCH REGULATORY REQUIREMENTS

Legal requirements may affect which conservation measures can be implemented in a service area. A utility may also have its own regulatory obligations that can be fulfilled through the use of conservation. Applicable local, state, and federal regulations should be researched as part of the program development.

Consider the following (Maddaus 1987a)

- Federal agency programs and activities
- State statutes and administrative codes for water use and water supply, including
  - Water rights law
  - Environmental permits
  - Water and energy programs
  - Building and plumbing codes
- Interstate agreements, court decrees, and regional water agreements

- Local ordinances and programs, including
  - Water use ordinances and regulations
  - Rate structures and policies
  - Land use planning and subdivision approval procedures
  - Local building and plumbing codes

Some of these regulatory requirements may complement certain conservation measures and make them easier to implement, while others may interfere with their use. Appendix A lists some of the agency contacts likely to be involved in conservation. If state agencies are not listed, local state government division of water supply can be contacted.

ESTABLISH CONSERVATION GOALS

Using utility data collected in Chapter 2, information on regulatory requirements and early surveys of public opinion, answer the following questions. These will enable the utility to establish conservation goals and select measures that best fulfill these goals (Maddaus 1987a).

- If there is a water supply shortage, is it limited to one portion of the service area, or is it a system-wide shortage?
- Is the supply shortage primarily short-term (drought, emergency shortage) or long-term (more than one year)? (Long-term supply shortages are the focus of this guidebook.)
- Is the shortage current or is it projected to occur in the future?
- What is the primary cause of the long-term supply shortage? Possibilities could include system leaks, inadequate water rights, pipeline delivery limitations, or inadequate water supply.
- Does the supply shortage occur during peak demand periods each day, during high water-use seasons of the year, or is it spread throughout the year?
- Is a small, medium, or large use reduction needed? Typically, a 1–10 percent reduction is considered small, 10–20 percent is medium, and more than 20 percent is large.
- When should these savings be achieved?
- Does there need to be a reduction in water use to meet state or federal regulatory requirements?
- Does the utility want to reduce water use in response to public or environmental concerns? If so, what sectors should be targeted?
- Information from public discussion, and the initial evaluation of potential conservation measures (from Chapters 3, 4, and 5) and answers to the previous questions should be used to articulate specific goals.

Example goals that can be used alone or in combination are

- Increase public awareness of conservation methods, and encourage utility customers to undertake these measures voluntarily. Contact _____ percent of all customers by the year ________.
- Decrease water use of existing customers by (1) reducing landscape water use and (2) retrofitting existing dwellings and commercial facilities with water-saving fixtures. Reduce existing customer residential water use by ___ percent by the year ________.

- Reduce the water use of new customers by encouraging efficient indoor and outdoor water use. Reduce new customer residential water use by ___ percent by the year ________ over similar existing homes.
- Reduce peak-day water use by ___ percent by the year ________ by focusing on landscape water use reduction.

SELECT PLAN MEASURES

Figure 6-1 presents an overview of measure selection based on utility characteristics. Among the customer measures profiled in Chapter 4, those that are the easiest to apply, meet goals at minimum cost, and indicate the greatest potential for savings in a service area should be selected. Worksheet 3 in Chapter 5, Evaluate Conservation Potential, can be used to estimate savings for different measures. Online calculation tools may be available from the state or from the Alliance for Water Efficiency, www.allianceforwaterefficiency.org.

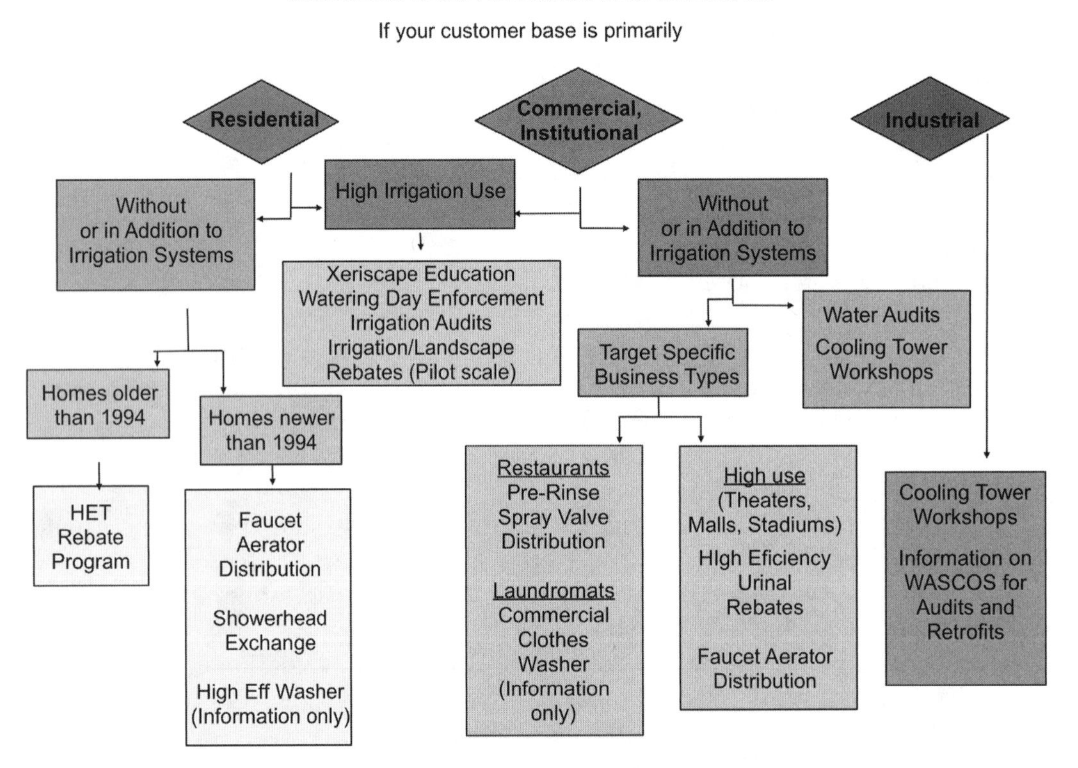

Figure 6-1. Overview of water conservation measure selection based on customer base. Measures outlined are those appropriate for small and medium-sized utilities with limited budgets. (Graphic: Water Media Services and Liz Block)

Applicable, Feasible, and Acceptable

In evaluating conservation measures, those that are applicable to the service area should be focused on. Landscaping conservation measures may not be appropriate if there is no significant increase in water use from irrigation during summer months or if outdoor use has not been identified as a large percentage of the customer water use.

If a goal is to reduce wastewater inflows to an overburdened treatment plant, landscape conservation measures may be ineffective if the system is served by a separate stormwater collection system.

The conservation measures, if any, are already in place in the service area should be considered. Conservation measures that exist in only some localities may be readily extended to cover the entire service area.

Feasibility within a utility water conservation budget is also critical. Assisting customers with retrofits may be initially beyond the financial capabilities of the utility, but with an increase in the water conservation budget, these programs may later be feasible.

Measures must also be acceptable to the utility's customers; otherwise, their success rate will be low. Utility staff can use their knowledge of the demographics and historical water use habits of the area to eliminate potentially unacceptable measures. The methods discussed in Chapters 3 and 4 are commonly used in various parts of the US and have not been found to be overly intrusive.

Acceptable Noneconomic Impacts

The changes required to reduce water use may have noneconomic impacts, which can be categorized as social/political, environmental, technical, and customer/public. These impacts are more difficult to quantify than economic impacts, but they may be as significant. Noneconomic impacts of the conservation measures discussed in this guidebook are shown in Table 6-1.

To evaluate noneconomic impacts for a service area, comprehensive lists of possible impacts should be compiled, and a decision made on whether a particular measure has positive, negative, or no impact. The most important considerations should be identified, and these issues brought up for public discussion. It should be noted that while some concerns cannot be measured or assigned a value, they may still be very important to the people whose support is being sought. A noneconomic negative impact may be significant enough to stop a program that has an otherwise favorable benefit–cost ratio, as evidenced by the growing number of engineering projects that are halted because of environmental concerns.

ESTIMATE BENEFITS AND COSTS

To have a feasible conservation plan, the total positive effects (benefits) of the plan must be greater than the total negative effects (costs). The greater the savings and the smaller the costs of the measures, the more economically attractive they will be.

A detailed benefit–cost analysis will permit a utility to compare the value of demand reduction measures (through conservation) with supply enhancement

Table 6-1. Noneconomic Impacts of Conservation Measures

	Toilet rebates	Industrial Water Audits	Landscape Water Audits	Landscaping Ordinance	Public Education
Impact					
Environmental/Technical					
New source development postponed or reduced	**Pos**	**Pos**	**Pos**	**Pos**	**Pos**
Reduced homeowner energy consumption	**Pos**	**Pos**			
Reduced utility energy consumption	**Pos**	**Pos**	**Pos**	**Pos**	
Increased life of water and wastewater treatment facilities	**Pos**	**Pos**	**Pos**	**Pos**	**Pos**
Increased streamflows	**Pos**	**Pos**	**Pos**	**Pos**	
Social/Political					
Create new jobs locally	**Pos**	**Pos**	**Pos**	**Pos**	
User and special-interest group opposition to program		**Neg**		**Neg**	
Requires mandatory ordinances					
Cooperation of enforcement authority to implement program may be difficult				**Neg**	
Cooperation with school department and other community departments may be difficult			**Neg**		**Neg**
Fairness of measure					
Requires landscaping attitude change					
Customer costs not equally shared between existing and new customers					**Neg**
Costs not equally shared between customers classes			**Neg**	**Neg**	**Neg**
Users who conserve will have lower energy bills		**Pos**	**Pos**	**Pos**	
Health and safety					**Pos**
Significant customer expense if mandatory		**Neg**			**Neg**
Source: Planning and Management Consultants, Ltd., et al. *Evaluating Urban Water Conservation Programs: A Procedures Manual*. Report prepared for the California Urban Water Agencies, Sacramento, Calif., 1992.					
Pos = positive impact.					
Neg = negative impact.					

measures (such as increased system capacity or other structural solutions). Planners can use the analysis to measure the impact of conservation on the capital facilities program to plan for future facilities requirements more accurately.

The AWWA M52 *Water Conservation Programs—A Planning Manual* has detailed instructions on how to prepare a benefit–cost analysis. Computer models can allow a utility to model benefit–cost analysis using net present value and benefit-to-cost ratio as economic indicators. In these models, benefit–cost analysis is performed from various perspectives including the utility and community (utility plus customer). Examples are the Water Conservation Savings Tracking Tool by the Alliance for Water Efficiency (www.allianceforwaterefficiency.org), the Demand Side Management Least Cost Planning Decision Support System or DSS model by Maddaus Water Management (Levin et al. 2006 or Maddaus and Maddaus 2006), or the IWR-MAIN model of CDM (www.iwrmain.com).

Estimating benefits and costs is helpful for informed decision-making, especially when a large budget increase is requested. However, it is not necessary in order to have a successful water conservation program and will be beyond the capacity of a small and many medium-sized utilities. A more informal assessment of economic impacts may be adequate to make comparisons and decisions, as in the following section. Meanwhile, the utility can keep track of certain information that can later be used in a more formal analysis. In particular, costs of the water conservation program, including administrative costs, can be calculated and should be mentioned.

Smaller utilities will find it easier just to calculate the cost of water saved, described in the following section, and select plan measures based on comparing these costs.

Even for a small utility, program cost information will be required to some extent during the public discussion and review period. An informal assessment of economic impacts can be made by comparing the program with another utility's program that is successful in similar achieving goals. Program budgets can be scaled using a dollars-per-person cost. This approach works when the size of the utilities used for comparison is not too different.

For example, if a nearby utility with 5,000 connections has a successful public education program similar to one a utility would like to implement for a system of 7,500 connections, and that utility's program cost is $10,000 per year, a budget about (7,500/5,000) × $10,000 = $15,000 per year should be implemented.

Typical Benefits and Costs

Benefits to the utility result from both short-term and long-term savings.

- Short-term savings are those that are not related to capital facilities and tend to result immediately from conservation activities. These include the cost savings from the reduced purchase of water, reduced costs of treatment chemicals, energy, and labor and materials required to handle reduced water production.
- Long-term savings are those associated with capital facilities (i.e., decreased cost for water and wastewater facilities because of reduced demand) or reduced water purchases.

The costs of water conservation programs fall into two main categories.

1. Cost for program implementation itself (carried by the utility and sometimes by the customer), which includes staff time, cost for hardware and public information materials, and the cost of any monetary incentives that may be offered.
2. Cost to the utility of reduced revenues resulting from decreased demand.

Other costs could include increased staff time for other municipal departments, such as planning or parks departments responsible for overseeing ordinances related to landscape water use or water-efficient landscaping.

Program Implementation Costs

Utility Costs

Costs to the utility can be expressed as follows, considering both in-house staff costs and contracted costs (where a private contractor performs some of the work).

- **In-House Cost** = Administration cost
  + Field labor hours × labor hourly rate, including overhead
  + Unit costs × number of units
  + Publicity cost
  + Evaluation/follow-up
- **Contracted Cost** = Administration cost
  + Number of sites × unit cost per site; includes program unit costs
  + Publicity cost
  + Evaluation/follow-up

Administration Costs

This is the staff time required to oversee staff, the work of consultants, or contracted field labor. Administration costs will be higher for new programs and for large consultant contracts. Administration costs are typically 10–15 percent of total program costs (Planning and Management Consultants, Ltd. et al. 1992).

Field Labor Costs

In addition to administrative staff time, the utility must supply labor to perform conservation work in the field. Field activities include water audits, leak repair and fixture installation, follow-up site visits, and door-to-door canvassing.

- Typically, trained utility personnel can perform 5–6 residential indoor water audits in one day. Only 3–4 residential indoor plus outdoor audits would typically be performed per day, depending on the detail required. Audit time includes report writing.
- Commercial/industrial water audits may be more time consuming and will require more specialized knowledge on the part of the auditor. Each audit may require a few days to a week or more, and costs will vary, depending on the complexity of the site.
- After base labor is determined for a commercial/industrial site, allow a factor of safety for items such as missed appointments, data interpretation, and follow-up visits. Unanticipated labor costs can double the original labor estimate, particularly for new programs.

Unit Costs

Many measures can be estimated on a unit cost basis or as cost per participant. Examples include retrofit kits, water audit programs, and rebate programs. Small programs typically have higher unit costs than larger programs because of the smaller number of participants.

Publicity Costs

All conservation programs should contain a public information element. Vehicles for public information include radio and television spots, local newspaper advertisements, flyers and bill stuffers, billboard advertising, workshops and seminars, and special demonstrations (for example, booths at community events). Larger utilities often employ public relations professionals to handle this aspect of their conservation program for maximum effect, but this is not necessary for smaller programs. Costs will be roughly proportional to the number of customers contacted.

Evaluation and Follow-up Costs

Typically, two types of follow-up need to be performed. The utility must keep records of the impact the conservation measure(s) is having (i.e., measure the water savings). The utility should also monitor how well the measure(s) is performing and whether program goals are being achieved. Chapter 7 discusses how to conduct these follow-up activities. Costs associated with follow-up activities may include the staff time needed to calculate water savings, and the costs of conducting public surveys to measure customer participation and satisfaction.

The best source of information for these costs is the experience of utilities that have run similar programs. Costs can be expressed on a unit basis (for example, $/dwelling unit or $/audit) and then transferred to a service area. A small utility should increase cost estimates because it will not have the economies of scale of larger programs.

Worksheet 4 calculates the annual in-house costs for conservation measures.

WORKSHEET 4
Determine Annual Costs (In-House)

MEASURE DESCRIPTION: ________________________________

Administration Costs		
1. Staff hours to administer the measure	______________	hrs/yr
2. Staff hourly rate, including overhead	$______________	/hr
3. **Administration costs** (Line 1 × Line 2)	$______________	/yr
Field Labor Costs		
4. Field labor hours	______________	hrs/yr
5. Field labor hourly rate, including overhead	$______________	/hr
6. **Field labor cost** (Line 4 × Line 5)	$______________	/yr
Unit Costs		
7. Unit cost of materials (i.e., retrofit kits, lawn watering kits)	$______________	/unit
8. Number of units distributed	______________	/yr
9. **Total materials cost** (Line 7 × Line 8)	$______________	/yr
Publicity Costs		
10. Materials cost (i.e., printing, outside services)	$______________	/yr
11. Advertising cost (i.e., newspaper, radio, TV)	$______________	/yr
12. **Total publicity costs** (Line 10 + Line 11)	$______________	/yr
Evaluation and Follow-up Costs		
13. **Labor costs**	$______________	/yr
Total Costs		
(Line 3 + Line 6 + Line 9 + Line 12 + Line 13)	$______________	/yr

Costs of Decreased Water Revenues

Decreases in water revenues resulting from conservation typically are small and occur over a long time period, allowing the utility to incorporate these changes into budget forecasts. Reductions in water revenues may be 0.5 to 2 percent per year over the life of the program, which is typically less than inflation in other utility costs. Cost savings from the short-term benefits of conservation (reduced energy, chemical, and treatment costs) will help offset these revenue decreases. Periodic rate increases can recover the lost revenue so that the actual cost will be insignificant.

CONSIDER OTHER PERSPECTIVES ON BENEFITS AND COSTS

It is also important to evaluate benefits and costs from other perspectives, to determine how willing other sectors might be to participate in the program. The most obvious group to consider is the customers, who will be either voluntary or involuntary participants in the program. Be aware that this evaluation may point to different alternatives than were suggested by the utility benefit–cost analysis.

Customer Benefits and Costs

Most conservation programs will result in some cost to the utility customer. Customer costs result from the purchase of special fixtures or other water-saving devices. The purpose of utility rebates and incentive programs is to offset the purchase and installation costs of these materials. Obviously, if customers' costs are too high, they will be reluctant to participate. Increased costs can also occur in commercial or industrial facilities where the installation of water-saving equipment requires additional operation and maintenance expenditures.

Benefits to the customer will result from reduced utility bills for water, wastewater, and energy. If the measure has a favorable benefit–cost ratio, the customer is more likely to implement it. In addition, secondary impacts on wastewater utilities may occur (reduced wastewater flows, lower cost of treatment, and cost savings from delayed facility construction). The noneconomic impacts of potential conservation measures is included in Table 6-1 "Measures With Acceptable Noneconomic Impacts." For detailed methodology on calculating customer benefits and costs, see the AWWA *M52 Water Conservation Programs—A Planning Manual* (Maddaus et al. 2006).

DEVELOP ALTERNATIVE APPROACHES

The conservation measures that were described in chapters 3 and 4 can be grouped into three general categories:

- Technological changes, the use of water-saving fixtures and hardware, both at the utility and customer level.
- Behavioral (voluntary) measures, such as low water-use landscaping and changes in irrigation practices. These measures could also be targeted (for example, voluntary for existing customers and mandatory for new customers).
- Changes in pricing structure to reduce demand.

In designing a conservation plan, a mixture of approaches should be selected to build flexibility into the program and prevent too much dependence on any one technique.

Three alternative conservation plans should be developed for internal and external review. This approach will encourage discussion and increase understanding of the choices that must be made. This type of dialogue is useful because the utility's governing board or regulatory agency and the public must support the program if it is to work.

Three levels of conservation could include

- Status quo (conservative) illustrates projected demand and estimate future capital facilities and operating costs without any additional conservation in place. This scenario can be produced by simply estimating future water usage, based on current water use practices and future growth (population, economic development, etc.).
- Stage 1 plan (preferred plan) where water would be saved using measures that are acceptable to the public. This plan may be the preferred option (and would be the minimum, moderate, or lower budget ranges of the maximum plan for which budget ranges have been specified earlier on Checklist 1).
- Stage 2 plan (aggressive plan) results in large water savings but might also carry the risk of strong public opposition or might be too expensive for a small or medium-sized utility (this would be the upper range budget range of the maximum plan for which budget ranges have been specified on Checklist 1).

In a practice called Conservation Measure Packaging (Maddaus et al. 2004), several measures can be bundled together starting with the most cost-effective until the overall program conservation targets or goals are met.

SELECT A PLAN

At this point, the utility has

- Researched regulatory requirements.
- Established goals.
- Identified measures that achieve water savings goals, are appropriate for the system, and have favorable noneconomic impacts.
- Made an informal assessment of economic impacts.
- Obtained public input during the planning process.

A conservation plan can be drafted to present to the public and decision makers. As discussed previously, the utility might want to present more than one level of plan for discussion (for example, a preferred option, plus a more conservative and a more aggressive approach). The draft plan should be modified based on public comments, in order to gain maximum public support for the program. The final plan should be resubmitted to the review group(s) for final approval.

Appendix D presents an outline for a typical water conservation plan (Maddaus 2006).

Implementing a Conservation Program

SUMMARY:

- *Discusses the process to put a program into place*

An implementation program should include program goals, a schedule, staff responsibilities, and a budget.

RESPONSIBILITIES OF THE CONSERVATION PROGRAM MANAGER

The water conservation program manager or water conservation coordinator will define the specific tasks to be carried out and determine the schedule and budget for each of these tasks. In a small utility, the conservation program manager will be the sole person performing most of these tasks and may work only part-time on conservation. For larger utilities, the water conservation program manager coordinates the overall program and has additional staff assigned to specific tasks, for example to coordinate residential landscape irrigation programs, CII programs, or education and public information programs.

IMPLEMENTATION TASKS

Specific tasks that could be performed as part of a conservation program include

- Coordinate with programs run by other agencies and neighboring utilities
- Disseminate information and conduct public relations activities
- Develop a public information and in-school education program
- Form and conduct a speakers' bureau program
- Oversee a utility leak reduction program
- Supervise retrofit device distribution

- Develop low water-use landscaping program
- Work with appropriate entities in a toilet rebate program
- Develop incentives to encourage conservation
- Revise local codes or ordinances to require water-saving fixtures

For each of these tasks, the program manager should prepare a description of staff responsibilities, estimate budget requirements, and determine a schedule. The cost of startup materials and staff training should be included in budget estimates.

OTHER PROGRAM PARTICIPANTS

In addition to the conservation program manager, other individuals and groups that may be involved in program implementation for a medium-sized utility include (Maddaus, 1987)

- **Water utility manager.** The utility manager approves the final conservation plan and authorizes budget and hiring requests. The utility manager will also extend formal requests for participation on a water conservation advisory committee if one is desired.
- **Water conservation advisory committee.** Small utilities or moderate-level programs generally do not have an advisory committee. If an advisory committee is used, its function is to advise the utility manager on conservation in the community
- **Consultant.** Consultants are sometimes used to determine water savings that result from conservation. They can also be employed to help launch a new program or expand an existing program.
- **Public information specialist.** This person can handle all aspects of the program relating to publicity and public relations. This task is generally handled in-house for a small utility.

PROGRAM MONITORING AND EVALUATION

Two kinds of program evaluations will be conducted 1) to determine actual water savings from the consumer program and 2) to identify ways to improve and refine the program over time.

Water Savings

The direct measurement of water savings is time consuming and may be difficult for a small utility to perform. See the AWWA M52 *Water Conservation Programs—A Planning Manual* (Maddaus et al. 2006), and *Handbook of Water Use and Conservation* (Vickers 2001) for detailed methodologies on evaluating water savings from water conservation measures and programs. An Excel-based tracking tool is also available from the Alliance for Water Efficiency (www.allianceforwaterefficiency.org).

Water use data should be saved before, during, and after implementation of a measure. This includes both raw utility pumping water data and customer billing data.

Cooperation with the utility's finance department is important to establish. At least a year of billing data before the measure and after the measure is needed. A billing system that allows data to be transferred to a spreadsheet is critical to facilitate this analysis.

Consumption data from the billing system should be compared for three periods.

1. Winter, to evaluate indoor use,
2. Summer, to evaluate outdoor use, and
3. Annual, to evaluate combined impact.

Various factors make before-and-after analysis of billing data difficult.

Possible variables for indoor water conservation measures, such as toilet retrofits, are

- Behavior: Did members of the household change water use practices?
- Other retrofits: Were other water conservation devices installed at the same time?
- Household size: Did the number of residents change?
- Inclusion of outdoor use: If outdoor use is included in the same water bill, were there any changes in outdoor use, including seasonal use changes?

Possible variables for outdoor water conservation measures are

- Behavior: Were there any changes in watering restrictions simultaneously in place?
- Behavior: Did the customer use the automatic irrigation system according to a preset schedule or turn the system on and off manually?
- Rainfall and temperature: Were there unusual weather conditions that would necessitate less or more water than average?
- Rainfall and temperature: Did a functioning rain sensor, SMS or ET controller, modify water use? If water savings from these devices are being studied, this question is of interest.

Important information to collect in evaluating program effectiveness, in addition to water savings, are

- Customer participation rate and customer satisfaction,
- Types and amount of customer contacts, and
- Description of problems and how they were overcome.

Other Estimates of Program Effectiveness and Durability of Savings

Another question is how long do savings last? Following up the initial evaluation of water savings after several years will allow the utility to investigate and address the question of persistence over time (City of Santa Cruz 2005).

Public surveys provide important information to measure participation rates and customer satisfaction. Customer surveys can also be used to collect specific data on water savings for use in calculating the overall impact of the program. A customer service questionnaire can be used to obtain feedback from customers on program satisfaction, to evaluate the usefulness of various components of the program, and to identify any areas or services in need of improvement. Two sample customer surveys—one general and one related to showerhead exchange—are included in Appendix C.

Telephone surveys have variable costs depending on the number of people contacted and the degree of data manipulation required. These surveys typically consist of 20 to 40 questions. Staff potentially could carry out small-scale surveys.

A mail survey for the same population is also possible. Postcard surveys with a mail-back feature have been used successfully by some utilities.

An email survey can be effective for participants who have provided an email address; however, some sectors of the customer base may not have access to email.

SUMMARY

If in the development of a conservation program the utility worked with a citizen's advisory group, citizen involvement should not stop once the conservation program has been implemented. A valuable network of informed supporters will have formed who can provide feedback and support.

Periodic workshops, seminars, and demonstrations can help keep the public interested and informed of conservation activities. Regular presentations to the commissioners or governing board are essential. The utility should also use the existing public network to modify and adapt the program to the public's needs and wants.

The utility should expect that the conservation program will require modification after the first year of operation. Ongoing efforts will increase the program's effectiveness and benefits over the years.

Program evaluators, including municipal leaders, will be looking for hard numbers on costs and participation. Accumulating data will provide requested information. With the documentation of successes, interest in the program and greater funding can be built, which will allow greater accomplishments through the program.

Appendix A

Contacts for Further Information

Federal

Federal agencies have information about pertinent federal legislation, such as the Energy Policy Act of 1992. They also have some public information materials and may have grant money available for research projects.

Federal Energy Management Program (FEMP), Water Efficiency
1000 Independence Avenue, SW
EE-2L
Washington, D.C. 20585-0121
Ph: 202-586-5772
http://www1.eere.energy.gov/femp/information/download_watergy.html

U.S. Army Corps of Engineers
Institute for Water Resources
7701 Telegraph Road
Alexandria, VA 22315
Ph: 703-428-8015
http://www.usace.army.mil/

U.S. Department of the Interior
Bureau of Reclamation
1849 C Street NW
Washington D.C. 20240-0001
Ph: 202-513-0501
http://www.usbr.gov/

U.S. Environmental Protection Agency
Office of Water (4101M)
1200 Pennsylvania Avenue, N.W.
Washington, D.C. 20460
Ph: 202-564-5700
http://www.epa.gov/water/

USEPA Water Sense Program
U.S. Environmental Protection Agency
Office of Wastewater Management (4204M)
1200 Pennsylvania Avenue, N.W.
Washington, D.C. 20460
Ph: 866-987-7367
http://www.epa.gov/watersense/

U.S. Census Bureau
4600 Silver Hill Road
Washington, D.C. 20233
Ph: 301-763-INFO (4636) or 800-923-8282
http://factfinder.census.gov/

U.S. Geological Survey(USGS)
Office of Water Information
5522 Research Park Drive
Baltimore, MD 21228
Ph: 443-498-5559
http://www.usgs.gov/

State

State agencies may require water conservation plans before certain permits or loans will be approved. Some have "how to" publications and public information materials available, and are a good source of information about other agencies that are implementing conservation.

Conservation activities in your state may be included within

- Department of Natural Resources
- Department of Water Resources
- Department of Health
- State Engineer's Office
- State Soil Conservation Service
- State Cooperative Extension Service
- State Environmental Protection Agency

A state's governor's office may also be able to help you find the appropriate agency.

Professional Organizations

Professional organizations offer a network of professionals working in the field. They have committees active in conservation, hold conferences, give papers, and can be a source of information about other agencies working in conservation.

American Rainwater Catchment Systems Association (ARCSA)
919 Congress Ave., Ste. 460
Austin, TX 78701
http://www.arcsa.org/

American Water Resources Association (AWRA)
PO Box 1626
Middleburg, VA 20118
Ph: 540-687-8390
http://www.awra.org/

American Water Works Association (AWWA)
6666 West Quincy Avenue
Denver, CO 80235
Ph: 303-794-7711
http://www.awwa.org

National Water Resources Association (NWRA)
3800 North Fairfax Drive, Suite 4
Arlington, VA 22203
Ph: 703-524-1544
http://www.nwra.org/

Soil and Water Conservation Society
945 SW Ankeny Road
Ankeny, IA 50023-9723
Ph: 515-289-2331
http://www.swcs.org/

Water Environment Federation
601 Wythe St.
Alexandria, VA 22314
Ph: 703-684-2400
http://www.weftec.com/

Other Sources of Information

Alliance for Water Efficiency
P.O. Box 804127
Chicago, IL 60680
Ph: 773-360-5100
http://allianceforwaterefficiency.org/

Association of California Water Agencies
910 K Street, Suite 100
Sacramento, CA 95814-3577
Ph: 916-441-4545
http://www.acwa.com

California Urban Water Conservation Council
455 Capitol Mall #703
Sacramento, CA 95814
Ph: 916-552-5885
http://www.cuwcc.com/

Center For Irrigation Technology
California State University, Fresno
5370 N. Chestnut Ave.
Fresno, CA 93740-8021
Ph: 559-278-2066
http://www.wateright.org/

Colorado Water Wise Council
P.O. Box 40202
Denver, CO 80204-0202
Ph: 303-893-2992
http://coloradowaterwise.org/

Conserve Florida Water Clearinghouse
Department of Environmental Engineering Sciences
A.P. Black Hall, Box 116450
University of Florida
Gainesville, FL 32611-6450
Ph: 352-392-7344
http://www.conservefloridawater.org

Consortium for Energy Efficiency (CEE)
98 North Washington St., Suite 101,
Boston, MA 02114-1918
Ph: 617-589-3949
http://www.cee1.org

Conservation Law Foundation
Massachusetts Advocacy Center
62 Summer Street
Boston, MA 02110-1016
Ph: 617-350-0990
http://www.clf.org/

Food Service Technology Center (FSTC)
Pacific Gas & Electric (PG&E)
12949 Alcosta Blvd. Suite 101
San Ramon, CA 94583
Ph: 925-866.2844
http://www.fishnick.com/

Fostering Sustainable Behavior
McKenzie-Mohr and Associates, Inc.
248 Eglinton Street
Fredericton, NB
Canada E3B-2W1
Ph: 506-455-5061
http://cbsm.com/

GreenPlumbers USA
4153 Northgate Blvd, Suite 1
Sacramento, CA 95834
Ph: 888-929-6207
http://www.greenplumbersusa.com/

Irrigation Association
6540 Arlington Boulevard
Falls Church, VA 22042-6638
Ph: 703-536-7080
http://www.irrigation.org/

Lane Community College Water Conservation Technician Program
Northwest Energy Education Institute
Lane Community College
4000 East 30th Avenue
Eugene, OR
Ph: 541-463-3977 or 800-769-9687
http://www.nweei.org/water-conservation-tech.html

Louisiana State University Green Laws Web Site
Louisiana Department of Agriculture & Forestry
P.O. Box 1628
Baton Rouge, LA 70821-1628
Ph: 225-925.4500
http://www.greenlaws.lsu.edu/

North American Weed Management Association
PO Box 687
300 Walnut St.
Meade, KS 67864
Ph: 620-873-8730
http://www.nawma.org

National Association of Exotic Pest Plant Councils
c/o University of Georgia
Center for Invasive Species and Ecosystem Health
P.O. Box 748
Tifton, GA 31793
Ph: 229-386-3298
http://www.naeppc.org

Pacific Institute
California Office
654 13th Street
Preservation Park
Oakland, CA 94612
Ph: 510-251-1600
http://www.pacinst.org

H2Ouse (a project of the California Urban Water Conservation Council)
http://www.h2ouse.org

Rocky Mountain Institute Water Program
2317 Snowmass Creek Road
Snowmass, CO 81654
Ph: 970-927-3851
http://www.rmi.org/

Smart Water Application Technologies
Irrigation Association
6540 Arlington Boulevard
Falls Church, VA 22042-6638
Ph: 703-536-7080
http://www.irrigation.org/SWAT/Industry/

Water Education Foundation (WEF)
717 K Street, Suite 517
Sacramento, CA 95814
Ph: 916-444-6240
http://www.water-ed.org

Waterwiser, AWWA Water Conservation Clearinghouse
http://www.waterwiser.org/

Local Government and Utility Websites

Arizona

Arizona Department of Water Resources
http://www.azwater.gov/dwr
Water Conservation Alliance of Southern Arizona (Water CASA)
Tucson
http://www.watercasa.org
Phoenix
http://www.ci.phoenix.az.us
http://www.amwua.org

California

California Department of Water Resources, Office of Water Efficiency
http://wwwowue.water.ca.gov
Monterey Peninsula Water Management District
http://www.mpwmd.dst.ca.us
Metropolitan Water District of Southern California
http://www.mwdh2o.com
http://www.bewaterwise.com
Municipal Water District of Orange County
http://www.mwdoc.com
East Bay Municipal District (EBMUD)
http://www.ebmud.com
Irvine Ranch Water District
http://www.irwd.com
Los Angeles Department of Water and Power
http://www.ladwp.com
North Marin Water District
http://www.nmwd.com
Santa Clara Valley Water District
http://www.scvwd.dst.ca.us
Santa Cruz
www.ci.santa-cruz.ca.us
San Diego
http://www.ci.san-diego.ca.us
http://www.sdcwa.org
San Francisco Public Utilities Commission
http://sfwater.org
San Jose Environmental Services Department
http://www.slowtheflow.com
Santa Monica
http://www01.smgov.net

Colorado

Aurora Water
http://www.aurorawater.org
Colorado Springs Utilities
http://www.csu.org
Denver Water
http://www.denverwater.org
Lower Colorado Region Water Conservation
http://www.usbr.gov

Florida

Tampa Bay Water
http://www.tampabaywater.org/
Miami-Dade Water and Sewer
http://www.miamidade.gov/conservation/
Pinellas County Utilities
http://www.pinellascounty.org/utilities/
Hillsborough County Water Resource Services
http://www.hillsboroughcounty.org

Georgia

Georgia Department of Natural Resources Pollution Prevention Assistance Division
http://www.p2ad.org
Environment Georgia
http://www.environmentgeorgia.org
WaterSmart, Cobb County, Ga.
http://water.cobbcountyga.gov/

Maryland

Washington Suburban Sanitary Commission
http://www.wssc.dst.md.us

Massachusetts

Massachusetts Water Resources Authority
http://www.mwra.state.ma.us
Executive Office of Energy & Environmental Affairs
http://www.mass.gov

Nevada

Southern Nevada Water Authority, Las Vegas
http://www.snwa.com

New Mexico

City of Santa Fe
http://www.water2conserve.com
City of Albuquerque
http://www.cabq.gov

New York

New York City Environment Agency
http://www.ci.nyc.ny.us

North Carolina

Town of Cary, N.C.
http://www.townofcary.org

Oregon

Portland Water Bureau
http://www.portlandonline.com
Regional Water Providers Consortium (Portland Area)
http://www.conserveh2o.org/

Pennsylvania

Philadelphia Water Department
http://www.phila.gov/water/

Texas

City of Austin
http://www.ci.austin.tx.us/watercon/
City of Dallas
http://savedallaswater.com/
City of Houston
http://www.publicworks.houstontx.gov
San Antonio Water System
http://www.saws.org/conservation/

Utah

Central Utah Water Conservancy District
http://www.cuwcd.com/

Virginia

Newport News Waterworks
http://www.nngov.com

Washington

Seattle Public Utilities
http://www.seattle.gov/util/Services/Water/index.asp
Saving Water Partnership
http://www.savingwater.org

International

City of Vancouver, British Columbia, Canada, Engineering Services
http://vancouver.ca
Environment Agency, U.K.
http://www.environment-agency.gov.uk
Environmental Canada, Canada
http://www.ec.gc.ca
Essex Suffolk Water, U.K.
http://www.eswater.co.uk/
METRON - Metropolitan Areas and Sustainable Use of Water, EEC
http://www.p2pays.org
North West Water, U.K.
http://www.nww.co.uk
Ottawa, Ontario, Canada
http://city.ottawa.on.ca
Toronto, Ontario, Canada
http://www.city.toronto.on.ca
Singapore Public Utilities Board, Singapore
http://www.pub.gov.sg
Water Services Department, Hong Kong, HKSAR China
http://www.wsd.gov.hk
Yorkshire Water, U.K.
http://www.yorkshirewater.com

Appendix B

CONDUCTING A RESIDENTIAL WATER AUDIT

An audit allows the utility to assist customers to understand ways to improve water efficiency, to distribute conservation information, and install devices. At the same time, the audit accumulates information on how water is used in the service area. Audits that include the installation of conservation devices provide the utility with a more accurate record of the number of devices installed, as opposed to customer pickup of devices in which there is no guarantee of installation.

When publicizing availability of the audits, the utility should emphasize the potential for lower water bills as a result of the audit. Promotion of the free items will interest the public.

Staff should conduct the audits in teams of two or more and should have a set route based on appointments made with customers. Staff should carry identification cards and show them to customers.

Audit staff should always explain the tasks they will be performing before they begin. They should also ask the customer's permission before altering any fixtures. If the customer wishes to follow the auditor throughout the procedure, the auditor has the opportunity to convey more information and should have information prepared on fixture flow rates, typical use in the area, etc.

The following outline suggests the types of activities that are performed as part of a residential audit. The actual content of an audit will depend on the other conservation measures the utility is pursuing. Audit forms and more detailed procedures, suitable for data entry of audit data, are found in Vickers 2001.

Service Meter

(Optional if a program of regular maintenance is already in place)

- Calibration/Flow Test
- Leak Test: To test if the leak is inside the house or outside the house, the main shut-off valve for water to the home may have to be turned off.
  - Ask the customer to turn off all water-using appliances in the home.

– Check the meter dial. If it is still moving, there is a leak in the service line that should be repaired promptly.

Indoor

- Bathroom
  - Toilets
    - Check for leaks.
      - Place a dye tablet or a few drops of food coloring in the tank. Do not flush the toilet. Do this ideally at the beginning of the indoor portion of the audit, and check back.
      - After 15-20 minutes, look in the bowl. If colored water is present, there is a leak.
      - Clean or replace the flapper that controls the entrance of water from the tank to the bowl. If the flapper needs to be replaced, the customer should be told that buying the correct replacement flapper is critical for water savings, and that correct model numbers are available on the following website: www.toiletflapper.org. Show them how to locate the toilet model number, generally in the tank.
      - Check the adjustment of the float arm. The ball on the end of the float arm should touch the surface of the water. No water should be running down the overflow tube.
    - Check the volume of the toilet using one of several methods:
      - If it is a 1.6 gpf (post-1994 toilet), the volume will be marked at the back of the seat. If it is an older toilet, the volume may be labeled inside the tank.
      - The volume can be measured by first turning off the water, flushing to empty the tank, and then refilling it from a bucket with a measured amount of water. Fill to the water stain line in the tank.
    - Offer to install toilet retrofit devices, or offer information on rebates for low-flow toilets if appropriate.
- Shower
  - Check shower flow rate. Using a watch with a second hand or a stopwatch, time how many seconds it takes to fill a 1-gallon jug or heavy plastic gallon bag. Some vendors of water conservation products sell specially marked plastic bags for doing flow tests.
  - Divide 60 seconds by the number of seconds it took to fill the jug or bag to the 1 gallon mark. For example, if it took 15 seconds to fill the jug, divide 60 by 15. The rate of flow is 4 gpm.
  - A pre-1994 showerhead will typically use over 3 gpm. Low flow showerheads required under EPAct 1992 use only 2.5 gpm.
  - If the flow rate is greater than 2.5 gpm, suggest possible savings by replacing the showerhead. If the utility has purchased showerheads in bulk, offer to install one. See Chapter 4 and the WaterSense website for discussion on choice of showerheads.

- Discuss with the customer the value of shorter showers.
  - Check for drips and leaks, and note these for repair by the customer. Once a new showerhead is installed, Teflon tape may prevent leaks.
- Lavatory Faucets
  - Check for drips and leaks.
  - Check sink faucet flow rate, as previously described.
  - Offer to install an aerator if one is not already in place.
  - Explain to the customer that 2.2 gpm has been the maximum flow rate allowed by code since 1994, but that faucet aerators with 0.5 gpm have been well received by customers and are now certified by WaterSense.
  - Offer to install a WaterSense-rated low-flow aerator. Use pliers and a rag to prevent scratching the faucet when removing the old aerator.
- Kitchen
  - Check for drips and leaks in the fixture.
  - Check the rate of flow from the faucet.
    - Use the method described for showers.
    - Explain to the customer that the maximum flow rate for new faucets since 1994 (EPAct 1992) is 2.5 gpm at 60 psi for residential faucets, but that lower flow rates have been well received.
  - Offer to install a low-flow aerator for the customer if one is not already in place.

Outdoor

- **Nonautomatic sprinkler system**

  (Audit conducted during irrigation season)
  - Check for leaks in the sprinkler, hose, or sprinkler system.
  - Check the position of the sprinklers. Determine whether the sprinklers cover only the area to be watered or whether the sprinklers need adjustment to prevent water from falling on homes, sidewalks, and other areas.
  - Provide the utility's lawn watering guide and instruct the customer on how to figure the length of time necessary for water to reach the root zone of lawns.
    - Show the customer how to identify the root zone.
    - Inform the customer that water should be applied slowly (to prevent runoff) until the water has reached the bottom of the root zone.
    - Instruct the customer to time how long it takes for water to reach the roots.
  - Help the customer develop a watering schedule in which the following items are identified:
    - Any restrictions on watering imposed by the local government.
    - The best time of day to water—either early morning or after the sun has gone down.
    - The numbers of days each week to water.
      - Water only when the lawn does not spring back when stepped on.
      - Apply water to reach the root zone once a week for temperate climates.
    - The length of time to water. This will depend on how long it takes for the customer's sprinkler to apply the correct amount of water to penetrate the root zone.

- Inform customers about low water-use landscaping (which plants use little water, where to buy such plants, and how to group plants on the site).
- Measure the flow rate of the sprinklers with catch cans (optional).

Adapted from: Washington Department of Ecology 1991.

- Automatic sprinkler system

  The following is a quick audit focused on timer (controller) settings. If a more in-depth irrigation system audit is desired, follow the audit procedure of the Irrigation Association. Information is available at www.irrigation.org. The latter includes calculation of distribution uniformity using a catch can test.
  - Check the timer first.
    - Is the current date and time correct? Are watering days set for 2 days/week summer and 1 day/week winter or whatever local watering restrictions allow? Or is the timer set using the controller's water budget feature?
    - Check the start time to make sure it complies with watering hours. For a timer with multiple start times, each start will run all the zones (customers may not understand that they need only one start time). Make sure it is set to run only one program. The watering days and start times on all other programs should be off. Multiple start times are good for clay soils that absorb poorly but otherwise not generally used for residences.
    - Go through each zone observing the coverage provided by the spray equipment.
      - Note if zones include both high water use and low water use plants.
      - Note if the zones contain sprays and rotors.
    - Check for overspray and adjust sprinklers. Note broken sprinklers. Note sprinklers overgrown by shrubbery and those spraying on the building.
    - If the yard slopes, but there are no check valves on final heads in a zone, recommend their installation.
    - If a rain sensor is required in the service area, note its presence and if present, check that it is functioning. While the sprinklers are going, use a ladder to climb to the sensor and press down on the top (in many models). After a short while, the sprinklers should turn off. Climbing to the rain sensor allows the setting to be checked. It should be set to turn off after no more than ½ in. of rainfall has been received. Newer models are preset for this amount. If there is no ladder available, spray the sensor with a hose and wait until it has received the preset amount; the sprinklers will turn off.
    - Return to the timer, and, if a catch can test will not be performed, set rotor zones for approximately 30 min. and spray zones for approximately 15 min. Explain to the customer that this setting is based on manufacturers' precipitation rates and a formula, that these run times put down ½ in. per application for typical spray or rotor equipment.
    - Ask the customer if he or she knows how to set and reset the timer for winter. If not, demonstrate and make sure he or she understands how to do it. The two methods are to reduce by a percentage using the timer's budget feature or to eliminate one day per week.
    - Make renovation suggestions based on separating high water use and low water use plants and sprays and rotors. The latter problem may be solved with installation of micro-rotors to replace spray heads in mixed zones.

Appendix C

EXAMPLE OF PUBLIC INFORMATION MATERIALS

- Sample Press Release
- Information on bill insert
- Sample tip sheets
- Informative billing
- Water Conservation Surveys

SAMPLE PRESS RELEASE

For Immediate Release
Contact:_____________
Date :_____________

Start Summer With Wise Water Use

Annual water use peaks between Memorial Day and Labor Day. Even though the _____________ area starts each summer with full reservoirs, the high demands of summer can bring supplies to a very low level before the reservoirs fill again with fall rains.

"The best way to be sure we have enough water to last all summer," notes ___________, "is to use it carefully from the beginning. We don't want to alarm our residents—we may have plenty of water. But we simply can't be sure until it rains in the fall. Last year we had record-breaking heat in October. If summer had ended after Labor Day, we would have had enough water. Having that extra two months of summer was just too much for our reserves. The best way to deal with the uncertainty is to assume we might have a long, hot summer and use water carefully."

One of the most effective water-saving practices relates to turf care. Most established lawns need no more than one inch of water a week. While each sprinkler system varies, residents can easily determine the amount of time it takes to put one inch of water on their lawn. Place an empty tuna can on the lawn before starting the sprinkler. If it takes 20 minutes to fill the can, then water the lawn 20 minutes each week.

Other lawn care tips include aerating lawns to be sure the water reaches the roots. Raising the mower up one notch also saves water. Even a little extra length provides a mulching effect to turf, helping it retain more moisture. Watering early in the morning saves water because little evaporates.

The __________________ (water provider) will be sharing water-saving tips with customers throughout the summer.

SAMPLE PRESS RELEASE

For Immediate Release
Contact:_____________
Date :_____________

"Skip a Week" During Cooler Months

The ________ water department is encouraging residents who irrigate their lawns to "skip a week" during the cooler months.

According to research by the University of _______, as temperatures grow cooler, grass doesn't need to be watered as often. Residents can use the following tips to determine when their grass needs water. Your lawn needs water when:

- Grass blades are folded in half
- Grass blades are blue-gray
- A footprint remains on the lawn

"The cooler months are a good time to try to gradually reduce your irrigation to train your lawn to need less water," said _______, ___________. "Over-watering makes your lawn less drought tolerant and can encourage pests. Using less water will encourage deeper grass and plant roots, which makes them more drought-tolerant."

Climatologists are predicting a La Niña event, which calls for a drier than normal fall, winter and spring.

The "skip a week" campaign builds on the department's landscaping public awareness campaign, which encourages water conservation during the late winter and spring when residents are thinking about gardening and improving their yards. For information about water restrictions, please visit the __ department's web site www.______ or call ______

(Modified from Southwest Florida Water Management District, 2008)

BILL INSERT TEXT, CITY OF ______ UTILITIES

Would you like to reduce your water bill by $85 dollars a year?

All you have to do is reduce your water usage by 1,000 gallons a month or 33 gallons day for your entire household. Most of these ways don't cost a cent.

1. **Fix toilet leaks!**
    - Toilet flushing and leaks account for approximately 30 percent of indoor water use.

- Toilets last a long time without leaking except for one weak link: the rubber flapper in the tank. Over time this rubber seal corrodes and does not close properly. Chemicals used to purify the water contribute to this corrosion and in-tank cleaning products containing harsh chemicals accelerate it.
- Sometimes the toilet "runs," and lets you know you have a leak. But sometimes the leak is silent.
- Check for silent leaks as follows:
  - Put a drop of food coloring in the tank. Do not flush.
  - Return 20 minutes later. Is there color in the bowl? If so you have a leak and need to replace the flapper.
- Replacing the flapper once a year is a preventative step. When you replace the flapper, take the original flapper to a plumbing supply store and ask for help choosing the correct replacement or call a plumber.
- If the toilet handle often sticks in the flush position, letting water run constantly, replace or adjust it.
- Sometimes the chain gets caught under the flapper. Jiggling the handle can fix that.
- For toilets not used on a regular basis, just turn the water intake valve off. Do this especially if you are going out of town for awhile.
- The amount saved by fixing toilet leaks is very variable, but *is easily over 1000 gallons per month.*

2. **Short showers are the easiest, cheapest way to save indoors.**
   - Showering uses about 17 percent of our indoor water.
   - A showerhead installed more than twelve years ago uses 3 gallons per minute or more, while a more recent showerhead uses 2.5 gallons per minute.
   - New low-flow showerheads are sold that use 2 gallons per minute or less, with air incorporated into the water flow.
   - The cheapest method to save is to just cut 5 minutes off your shower. Assuming a flow rate of 3 gallons per minute, you will save 15 gallons per shower. That is 450 gallons per month times the number of people in the household or *easily 1,000 gallons saved right there.*
   - Short showers also save on your energy bill, because your hot water heater doesn't have to work as long.
3. **Install faucet aerators**
   - Faucets, including bathroom and kitchen faucets, make up about 16 percent of typical indoor water use.
   - Faucet aerators are circular metal devices that screw into the faucet spout that mix air into the water. Basic bathroom faucet aerators sell for as little as $1.
   - Bathroom faucet aerators currently sold use 2.2 gallons per minute or less, while older bathroom sink faucets may use up to 5 gallons per minute.
   - If you add a faucet aerator, you could be saving 2.8 gallons each minute the faucet runs.
   - Periodically unscrewing the aerator and cleaning out grit and scale buildup is helpful to maintain a good flow.
   - If your faucet is dripping, washers may be worn and should be replaced. Follow directions in books or websites or call a plumber.

- Replacing an older faucet is estimated to save 7 gallons per household day or 210 gallons per month.
- Best of all, just be aware of how much water goes down the drain and turn off faucets when you are not actually using the water.
- By running hot or warm water a shorter time, you also save on your energy bill.

4. **Full loads with your clothes washer**
   - Clothes washing makes up about 22 percent of typical indoor water use.
   - It takes about 35 gallons per cycle with an older clothes washer of the top loading type (while new front-loading Energy Star clothes washers use less than half that amount and save energy at the same time).
   - What people think is the maximum load that the machine can handle is generally much less than what the machine can actually handle. Try filling the machine a little fuller until you are comfortable with what your machine can handle.
   - Eliminating one load per week with an older washer saves at least 35 gallons per week or 160 gallons per month. Washing fewer loads will also save significantly on your energy bill.
5. **Stop other household leaks.**
   - Leaks make up as much as 14 percent of typical indoor water use.
   - If you have an unusually high water bill, check your water meter. With all water use in the house turned off, if the reading changes or you see the little triangle or sweep hand turning, you have a leak! Find the leak yourself or call a plumber.
   - For outdoor faucets that leak, wrap Teflon tape around the threads of a hose bib.
   - Use a "hose gun" to help prevent hidden hoses from running and wasting water. Or just disconnect the hose from the hose bib when not in use.
6. **Tune up your irrigation system**
   - This won't reduce your water bill if you have a well, but it will save on energy to run the pump. And it helps save water in general.
   - Position sprinklers to keep water off paved areas.
   - Set your timer for the 2 day per week watering restrictions. One half inch per application is probably adequate.
   - In the cooler months, one day a week should be adequate or turn the system off altogether.
   - Make sure your rain sensor is functioning for the rainy months.
   - An irrigation technician can help you with timer settings and adjustments and may be available through the city at a reduced rate. Check the city website for current offerings.
   - Rain barrels can store rainwater to use in your garden. Sixty gallon rain barrels with screens and spigots are sold at _____, phone ______.

Following these tips can help you save money and assist the City of _____ lower costs to provide water to all its citizens.

SAMPLE TIP SHEETS

Do you have an older home built before 1985 that has not had the bathroom remodeled? Do you have pliers? Would you want a free device that will save your family money?

"How Can I Save Water Indoors?" you ask...

In the Bathroom:

- Replace your toilet with a high-efficiency toilet that uses only 1.28 gallons per flush. Older toilets may use as much as 5 gallons per flush.
- As you look for a new toilet, be sure to look for the WaterSense label.
- A heavy glass jar can help you save water easily, and at a low cost. An empty pickle jar will work great. Lift off the toilet tank lid and hold the jar in the water until it fills. Place the jar in the tank bottom, away from any moving parts. Put the tank lid back on. You'll be saving water with every flush.
- Once a year, check for toilet leaks. Lift off the toilet tank lid and drip 10 drops of food coloring into the tank. After 10–15 minutes, check for dye color in the toilet bowl—if you see any dye color, your toilet has a leak. Toilet leaks can waste hundreds of gallons a day.
- Turn the faucet off when brushing your teeth, shaving, or washing your face and you'll save 2–3 gallons of water each minute.
- Use faucets at low volume.
- Install high-efficiency, low-flow faucet aerators that use no more than 2.2 gallons of water per minute. Older faucets use between 3 and 7 gallons per minute.
- Fix leaky faucets right away. Simply replacing the washer may fix the leaks. Even small drips caused by worn washers can waste 20 gallons of water or more a day. Large leaks can waste hundreds of gallons.
- Shorten your shower time—older showerheads use over 4 gallons of water a minute.
- Install a high-efficiency, low-flow showerhead that uses no more than 2 gallons of water per minute.
- When taking a bath, fill the tub up halfway or less. A full tub can hold more than 50 gallons of water.

In the Kitchen:

- Instead of letting the faucet run until the water is cold enough to drink, keep a container of drinking water in the refrigerator. Older faucets can use from 3 to 7 gallons of water per minute.
- Clean vegetables and fruits in a pan of water, not under a running faucet. Use a vegetable brush to remove dirt.
- When defrosting food, thaw it in the refrigerator instead of under running water. This may take an extra day or two.
- For your kitchen sink, install a high-efficiency, low-flow faucet aerator that uses no more than 2.5 gallons of water per minute.

- If you wash dishes by hand, fill the sink or a pan with soapy water instead of letting the faucet run while soaping dishes. And don't let the faucet run while rinsing off dishes—rinse dishes in a filled sink or a pan of water.
- Run the dishwasher only when it's fully loaded. Older dishwashers use between 12 and 15 gallons of water, full or empty. When loading the dishwasher, scrape food off of dishes and pots instead of rinsing them.

Other Chores:

- Presoak clothes in the washing machine only when absolutely necessary.
- Run the washing machine only when it's fully loaded. Some washing machines have controls that let you select the load size. Older washing machines use over 35 gallons of water per full load.
- If you're thinking about buying a new washing machine, check out front-loading washers. These washers can cost more, but they use about half as much water per load as older top-loading machines.

Remember, use what you need and save the rest!

"How Can I Save Water Outdoors?" you ask...

- Help your soil to hold the right amount of water:
  - **Clay soil**: Add organic material such as compost or peat moss. Tilling or spading will help loosen the soil. Clay soil absorbs water very slowly, so water only as fast as the soil can absorb. Don't waste water by letting it run off.
  - **Sandy soil**: Water can run through sandy soil so quickly that plants don't have a chance to absorb it. Add organic material to supplement the soil and slow down water flow.
  - **Loamy soil**: Loamy soil is the best. It's a blend of sand, silt, and clay. It absorbs water readily and stores it for use by plants.
- When you water your lawn, water infrequently, but thoroughly so that moisture soaks down to the roots, about four to six inches. This encourages deeper, healthier root systems and allows the lawn to go without water for a longer time. Watering too frequently encourages the plants to have shallow roots.
- Water during the cool of the day—late evening, early morning. This helps reduce the amount of water that is wasted through evaporation. Morning is best, as night watering may encourage growth of molds.
- If you don't have an automatic sprinkling system, set a kitchen timer or invest in a sprinkler timer to help prevent overwatering. Outdoor faucets can flow at rates as high as 300 gallons per hour. A lot of water can be wasted in a short period of time if you forget to turn your sprinklers off.
- Weeds are water thieves and will rob your plants of water and nutrients. Try to keep your lawn and garden weed-free—spot spray or remove weeds as they appear.
- If it doesn't grow, don't water it. Position sprinklers so water doesn't land in the gutters or on any paved areas.

- Don't water on windy days. Water will go everywhere except where it is needed.
- Adjust your lawn mower to a higher setting and mow more frequently. Longer grass blades provide shade and help hold moisture in longer.
- Lay mulch around trees and plants at least 1 to 2 inches deep to retain moisture, slow evaporation, and discourage weed growth.
- Consider installing drip irrigation systems around trees and shrubs. Drip systems allow water to flow slowly to roots. This encourages strong root systems. Drip irrigation systems also reduce water loss from evaporation.
- Use a broom, not a hose, to clean driveways, sidewalks, and other hard surfaces.
- Wash your car with a bucket and sponge. Don't let the hose run while washing your car—and use a hose with a shut-off nozzle to rinse.
- Check for leaks in pipes, hoses, faucets, and couplings. Leaks can waste a lot of water.
- If you own a pool, get a cover for it to help prevent evaporation. An average sized pool can lose about 1,000 gallons of water per month. A pool cover can cut this loss by up to 90 percent.
- Remember, use what you need and save the rest!

INFORMATIVE BILLING

Informative billing could include:

- The utility's rate structure,
- Amount of water used in the current month,
- Number of days in the billing cycle if this varies,
- Amount of water used in the previous month,
- Amount of water used in the same month of the previous year,
- The average usage of all customers in the same customer class (single family residential, for example),
- Seasonal rates and/or drought rates and applicable months, or
- Other information deemed appropriate by the utility. For example, water bills could give information on how much a large use customer could save simply by reducing irrigation by one cycle per week.
- Educational information through water bill inserts or other means. For water systems where residential water use is greater than 100 gallons per capita per day, this should occur at least once a year.

Source (Green and Yingling 2007). See also Figures 3-4 and 3-5.

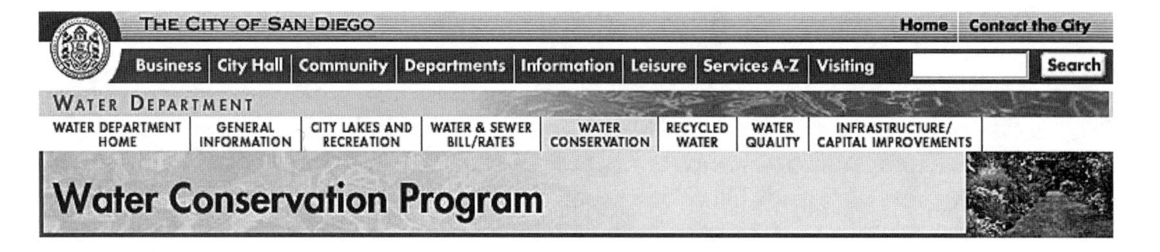

Water Conservation Survey

The City of San Diego's Water Conservation Program is continually looking for ways to improve our service to our customers. In order to get better, we need to hear from you! Please fill out this short survey about the ways you conserve water, and what you think of the programs and services we offer. In appreciation for your time and effort, you will receive a free *Southern California Heritage Gardening Guide* CD. Thanks for your help!

This survey must be completed by a person over the age of eighteen. Please verify by checking the box below.

☐ I am eighteen years old or older.

Please enter your 5-digit zip/postal code: [ ]

1. How often do you hear about water conservation in the media?

○ Daily ○ Weekly ○ Monthly ○ Bi-monthly

○ Other [ ]

2. How do you access water-saving tips and programs? (Check all that apply.)

☐ Hotline ☐ Water Dept. Newsletter ☐ Website

☐ Word of mouth ☐ Local Media ☐ Water Dept. Bill

☐ Other [ ]

3. What techniques have you used to save water? (Check all that apply.)

☐ Turn off water while brushing teeth/shaving
☐ Wash full loads of dishes or laundry
☐ Sweep debris off driveways, patios, etc.
☐ Collect cold water for other uses while waiting for water to heat up
☐ Other [ ]
☐ Take shorter showers (5 minutes or less)
☐ Adjust sprinkler schedules regularly
☐ Redirect downspouts towards lawn/plants
☐ Use California Native or Low Water-Use plants

4. How do you set your sprinkler schedule?

☐ According to the weather
☐ According to the Landscape Watering Calculator
☐ Landscape maintenance professional sets schedule
☐ Schedule remains the same all year long
☐ I have no irrigation system that I maintain
☐ Other [ ]

5. Do you plan to landscape your property in the near future? ○ Yes ○ No

If yes, for what type of property? ○ Residential ○ Commercial

Water Conservation Survey courtesy Water Department, City of San Diego

6. How often do you check your faucets/fixtures for leaks?

○ Monthly ○ Bi-monthly ○ Annually ○ Bi-annually

○ Never ○ Other [ ]

7. Have you participated in one of the City's Water Conservation programs?

○ Yes ○ No

(If YES, continue to question 8. If NO, skip to question 14.)

8. If yes, select a program below which you participated in most recently.

[--select one--]

9. Please rate the service you received:

	Exceptional	Good	Satisfactory	Needs Improvement
Friendliness of telephone operator?	○	○	○	○
Answers to technical questions?	○	○	○	○
Friendliness of program staff?	○	○	○	○
Professional appearance of staff?	○	○	○	○
Overall experience?	○	○	○	○

10. Would you recommend this program? ○ Yes ○ No

11. Have you participated in other programs? ○ Yes ○ No

12. If yes, which ones? (Check all that apply.)

- ☐ Residential Water Survey Program
- ☐ High-Efficiency Clothes Washer Voucher Program
- ☐ Ultra-Low Flush Toilet Voucher Program
- ☐ Landscape Watering Calculator
- ☐ Commercial, Industrial, and Institutional Water Conservation Survey Program
- ☐ Showerhead Retrofit Program
- ☐ Commercial Landscape Survey Program
- ☐ Commercial Landscape Incentive Program
- ☐ Plumbing Retrofit Upon Resale Ordinance

13. Are you interested in telling us your water conservation story to share with local media? ○ Yes ○ No

If YES, please check your level of participation:

○ Telephone interview ○ Written description of water conservation story below.

[ ]

14. Additional comments and ideas for how we can improve:

[ ]

Water Conservation Survey courtesy Water Department, City of San Diego

15. To receive information on Water Conservation Programs and Services, check all that apply:

- ☐ Residential Water Survey Program
- ☐ High-Efficiency Clothes Washer Voucher Program
- ☐ Ultra-Low Flush Toilet Voucher Program
- ☐ Landscape Watering Calculator
- ☐ Commercial, Industrial, and Institutional Water Conservation Survey Program
- ☐ Water Conservation Garden
- ☐ Commercial Landscape Survey Program
- ☐ Commercial Landscape Incentive Program
- ☐ Plumbing Retrofit Upon Resale Ordinance

16. To receive your FREE *Southern California Gardening Guide* CD please complete the following (Optional):

Name [ ] Address [ ]

City [ ] State [ ] Zip Code [ ]

Phone [ ] Email [ ]

17. A. **Would you like us to contact you in order to respond to any of your questions or concerns?** ○ Yes ○ No

B. **Would you like to be added to our distribution list to receive press releases and additional information on water conservation programs?** ○ Yes ○ No

View the "Survey Question of the Month" and our response.

Submit Cancel

For information about recycled water, visit the Recycled Water Program and Water Resuse Study web pages.

Water Conservation Survey courtesy Water Department, City of San Diego

Customer Name:____________________
Last *First*

Service Address:____________________
As it appears on your water bill

City:__________ State:____ Zip:______

Customer Phone: ____________________

NMB Account Number: ____________________
Optional *located on your water bill*

No Postage Necessary if Mailed in the United States

Myron Rosner, Mayor
Philippe Derose, Councilperson
John Patrick Julien, Councilperson
Barbara Kramer, Councilperson
Frantz Pierre, Councilperson
Phyllis S. Smith, Councilperson
Beth E. Spiegel, Councilperson
Kelvin L. Baker, City Manager
Darcee S. Siegel, City Attorney
Susan A. Owens, City Clerk

City of North Miami Beach
Public Services Department
Water Conservation Division
17050 N.E. 19th Avenue
North Miami Beach, FL 33162

(305) 948-2967
www.nmbworks.com

Indoor Water Saver Kit courtesy City of North Miami Beach

Leak Detection Tablets

Step 1 - Carefully remove top from toilet tank.

Step 2 - Drop two tablets into tank; DO NOT FLUSH.

Step 3 - Wait 10 Minutes.

Step 4 - If blue color appears in the bowl, you have a leak that is wasting water.

Showerhead Installation

Step 1 - Remove old showerhead from shower arm. If you need to, carefully use a wrench. Use a piece of cloth to protect finish.

Step 2 - Before installing new showerhead, turn on water to flush pipe.

Step 3 - Turn off water, wrap pipe threads with teflon tape. Screw on new showerhead and hand tighten.

Step 4 - Turn on water and test showerhead.
If it leaks, gently tighten with wrench.
DO NOT OVER TIGHTEN.

Faucet Aerator Installation

Step 1 - Remove old aerator, carefully use a wrench, if necessary.

Step 2 - Turn on water to flush faucet.

Step 3 - Install new faucet aerator using washers that are supplied.

Step 4 - Hand tighten first, carefully use wrench to tighten securely.
DO NOT OVER TIGHTEN.

Drip Gauge

Step 1 - Remove cap from drip gauge.

Step 2 - Hold drip gauge under dripping faucet or other leaking device for five (5) seconds and remove.

Step 3 - Referencing the scales on the side of the gauge, you can estimate how much water is being wasted through the drip or leak.

Step 4 - Repair all leaks as soon as possible.

SHOWERHEAD EXCHANGE PROGRAM SURVEY

We value your opinion. ***Please complete the survey below and place it in your mailbox. NO postage necessary. Your comments allow us to evaluate this program. Thank you!***

Circle the appropriate answers and return

1. How did you learn about the North Miami Beach Showerhead Exchange Program?	**Newspaper Ad**	**Bus Bench**	**Other** (*specify*)
2. Have you installed your new showerhead?	YES	NO	Not sure
3. Have you installed the faucet aerators included in your kit?	YES	NO	Not sure
4. Have you tested your toilet(s) for leaks using the dye tablets included in your kit?	YES	NO	Not sure
5. After using the dye tablets, did you find that your toilet(s) leaked?	YES	NO	Not sure
6. Did you find that any of your faucets leaked?	YES	NO	Not sure
7. Did you find that the devices were easy to install?	YES	NO	Not sure
8. Do you feel the devices and water conservation tips included in the kit are helping you save water?	YES	NO	Not sure
9. Do you feel the devices and water conservation tips included in the kit are helping you lower your water bill?	YES	NO	Not sure
10. Overall, are you satisfied with North Miami Beach's Water Saver Kit, its contents and Showerhead Exchange Program?	YES	NO	Not sure
11. Would you support other programs that promote water conservation?	YES	NO	Not sure

Additional comments:

Indoor Water Saver Kit courtesy City of North Miami Beach

Appendix D

TYPICAL WATER CONSERVATION PLAN OUTLINE (FROM MADDAUS 2006)

- INTRODUCTION AND SUMMARY
  - Purpose and scope of plan
  - Plan submittal requirement
  - Plan development and public participation
  - Plan elements
  - Resolution for adopting the plan
- STUDY AREA CHARACTERISTICS
  - History of water system
  - Demographic forecasts
- ANALYSIS OF HISTORICAL AND PROJECTED WATER DEMAND
  - Historical water use
  - Analysis of water use by customer group
  - Summary of historical and projected demand without conservation
  - Impact of new plumbing code on water use
- WATER SUPPLY
  - Sources of water
  - Groundwater
  - Surface water
  - Overall supply and demand balance
- RECLAIMED WATER PLAN
  - Results of previous studies
  - Plans for reuse
- CURRENT WATER CONSERVATION PROGRAM
  - Measures implemented by water wholesaler and the city
  - Management of nonrevenue water
- ALTERNATIVE WATER CONSERVATION MEASURES
  - List of conservation measures considered

- EVALUATION OF LONG-TERM WATER CONSERVATION MEASURES
  - Menu of water conservation alternative programs
  - Estimated water savings
  - Costs of measures
  - Results of benefit–cost analysis
- RECOMMENDED PLAN
  - Selection criteria
  - Description of recommended plan
  - Projected water savings
  - Benefits
  - Implementation schedule
  - Budget and staffing
- WATER SHORTAGE PLAN
  - Worst case water supply
  - Plan elements
  - Water use restrictions
  - Water supply emergency
  - Water rate structure

Appendix E

METHOD TO ESTIMATE WATER USAGE FROM AN UNMETERED SUPPLY

If your system is not metered and is from a pumped source, you can use the readings from your electric power supply meter to calculate water production, using the following formula.

Pumping Volume = Pumping Rate × Time of Operation

1. Calculate the water pumping rate by measuring pump output flow rate. (Measure time to fill a water tank and compute rate in gallons per minute.)
2. Calculate the time of pump operation from the electric meter reading. Procedure:
   a. Record the beginning and ending power meter dial readings for the period of interest (day, month, year). Figure the total kilowatt per hour (kWh) produced (final minus initial dial meter reading).
   b. Multiply the kWh supplied during the period times the scale factor printed on the face of the meter.
   c. Record the Kh factor from the meter face.
   d. During pump operation, time the rotation of the meter disc by counting revolutions (for more than 10 revolutions), and record the number of revolutions and the total time in seconds.
   e. Compute the instantaneous kilowatt demand with the following formula:

$$kW\ (inst) =$$

$$\frac{(\#\ revolutions \times Kh\ factor \times 3.6)}{(total\ time\ in\ seconds\ measured\ in\ Step\ d)}$$

   f. Compute the time of pump operation (in hours) with the following formula:

Pumping Time = Total kWh/kW (inst)

3. Multiply pumping rate (Step 1) by time of pump operation (Step 2) to arrive at volume of water per period (day, month, year).

Common Water Units
1 cubic foot (cf) = 7.48 gal
1 ccf (commonly used by water utilities as "1 unit") = 748 gal
An acre-foot is the volume covering an acre of area to a depth of one foot.
1 acre-foot = 325,851 gal or 43,560 cf
1 million gal/day = 3.07 acre-feet/day
To convert gal to acre-feet, divide by 325,851.43.
To convert acre-feet to gal, multiply by 325,851.43.

Appendix F

AWWA POLICY ON WATER CONSERVATION

Policy

Adopted by the Board of Directors Jan. 27, 1991, revised Jan. 31, 1993, and June 15, 1997, and reaffirmed Jan. 20, 2002.

The American Water Works Association (AWWA) strongly encourages water utilities to adopt policies and procedures that result in the efficient use of water, in their operations and by the public, through a balanced approach combining demand management and phased source development.

To this end, AWWA supports the following water conservation principles and practices:

1. Efficient utilization of sources of supply;
2. Appropriate facility rehabilitation or replacement;
3. Leak detection and repair;
4. Accurate monitoring of consumption and billing based on metered usage;
5. Full cost pricing;
6. Establishment of water-use-efficiency standards for new plumbing fixtures and appliances and the encouragement of conversion of existing high-water-use plumbing fixtures to more efficient designs;
7. Encouragement of the use of efficient irrigation systems and landscape materials;
8. Development and use of educational materials on water conservation;
9. Public information programs promoting efficient practices and water conservation by all customers;
10. Integrated resource planning;
11. Water reuse for appropriate uses; and
12. Continued research on efficient water use practices.

Glossary

Adapted plants — Plants not indigenous to an area but from a similar climate that require little or no supplemental irrigation once established.

Account — A connection to a water system, which is billed for service.

Acre-foot — A volume of water that would cover one acre to a depth of one foot, or 325,850 gallons of water.

Alliance for Water Efficiency (AWE) — A nonprofit membership organization promoting the efficient and sustainable use of water. www.allianceforwaterefficiency.org

American Water Works Association (AWWA) — A professional organization serving the drinking water supply profession.

Annual Average Demand — The amount of water required to meet demand in a typical year, usually expressed as millions of gallons per year or as cubic feet per year. Also termed *average demand* or *average annual water delivery*.

Apparent losses — In a distribution system water audit apparent losses represent the "paper" losses that occur when volumes of water reach a user, but are not properly measured or recorded. They include customer meter inaccuracies, unauthorized consumption and data handling error in customer billing systems. Apparent losses cause water utilities a loss of revenue but also interject a degree of error in the assessment of customer consumption, making it more difficult to evaluate the success of water conservation and loss control measures.

Arid climate — A climate characterized by less than 10 inches of annual precipitation.

As-built plans — Site plans reflecting the actual constructed conditions of a landscape irrigation system or other facility installation.

Audit (end-use) — A systematic accounting of water uses conducted to identify opportunities for improved efficiency.

Authorized unmetered water use — Unmetered water used for beneficial purposes (e.g., firefighting and training, flushing mains, etc.).

Automatic irrigation — Delivery of water to a landscape using a timer, a system of valves, and sprinklers.

Automatic irrigation controller — An irrigation timer capable of operating valve stations to set the days and length of time of water applications.

Average-day demand (ADD) — A water system's average daily use based on total annual water production divided by 365.

Avoided cost — The cost of an activity or facility that could be avoided by choosing an alternative course of action.

Ballcock — A float actuated valve, part of the toilet trim in the toilet tank that controls the refill water flowing into the toilet tank when it is not full.

Baseline — An established value or trend used for comparison when conditions are altered, (e.g., introduction of water conservation measures).

Benchmark — Expected performance of a best management practice or water conservation measure.

Benefit-cost analysis — A comparison of total benefits to total costs, usually expressed in monetary terms, used to measure efficiency and evaluate alternatives; see also *avoided cost.*

Best management practice (BMP) — A practice or combination of practices established as the most practicable means of increasing water use efficiency.

Bill stuffer — An advertisement or notice included with a utility bill.

Billing cycle — The regular interval of time when customers' meters are read and bills are issued, generally every month (monthly) or two months (bi-monthly).

Bleed-off — Draining off the water in a cooling tower reservoir to avoid the buildup of excess dissolved solids. Also referred to as *blowdown.*

Block — A quantity of water for which a price per unit of water (or billing rate) is established.

Block-rate pricing — A method of charging for water based on the volume used. As more water is used, the price increases (or decreases) through a series of blocks. These pricing structures are designed to encourage efficient use of a resource.

Blowdown — Draining off the water in a cooling tower reservoir to avoid the buildup of excess dissolved solids. Also referred to as *bleed-off.*

Bubbler — A type of sprinkler head that delivers a relatively large volume of water to a level area where standing water gradually infiltrates into the soil. The flow rate is large relative to the area to which the water is delivered. Bubblers are used to irrigate trees and shrubs.

Capacity buy — Process of "finding" water by retrofitting older homes and buildings.

Capita — Latin for *person.*

Capital Facilities — Physical facilities required for utility operation. These may include water storage, transmission, and treatment facilities.

Catch-can test — Measurement of a sprinkler system's application rate. Test involves placing graduated containers at evenly spaced intervals throughout an irrigated area and measuring the depth of water collected in the cans over a given period of time.

CCF — 100 cubic feet of water an amount equivalent to 748 gallons.

Central irrigation control system — A computerized system for programming irrigation controllers from a central location; using a personal computer and radio waves or hard wiring to send program information to geographically distant controllers.

Check valve — A device that prevents drainage of water down to the low points of an irrigation system after the system is shut off. Also called anti-drain valve. A valve that allows flow in only one direction, preventing backflow.

Cistern — A tank (often underground) used to store water (often rainwater or graywater).

CII — Commercial, Institutional, Industrial customer sector, also referred to as *ICI.*

Class — Customers having similar characteristics (commercial, single-family residential, etc.) grouped together for billing or program purposes.

Climate factor — Evapotranspiration minus precipitation. One of the four factors used to determine landscape water use.

Closed loop cooling tower — Water-conserving cooling tower system in which water used for cooling is recycled through a piping system that cools the water; the water is cooled as air exchanges heat with the pipes.

Commercial facilities — A customer category that includes retail businesses, restaurants, hotels, office buildings, and car washes.

Commercial user — Customers who use water at a place of business, such as hotels, restaurants, office buildings, commercial businesses or other places of commerce. These do not include multi-family residences, agricultural users, or customers that fall within the industrial or institutional classifications.

Commodity charge — See *variable charge.*

Commodity rate — Charging for water based on the volume of use. Not a flat or fixed rate.

Connection fee — A charge assessed to a new account by a water utility that generally covers the cost of hooking up to the system and compensates the utility for prior water system improvements that made the capacity available.

Conservation — Increasing the efficiency of energy use, water use, production, or distribution; the act of conserving or preserving from injury or loss; the protection of rivers, forests, and other natural resources.

Conservation coordinator — A water conservation measure where the water supplier designates a water conservation coordinator and support staff (if necessary).

Conservation pricing — Water rate structures that increase the price of water as more water is used with the goal of encouraging more efficient use.

Conservation rate structure — A pricing structure billed by the quantity of commodity delivered and tied to the costs associated with that delivery, designed to provide an accurate price signal to the consumer. An increasing block rate structure, if the top tier equals the utility's marginal cost of new water, is one example of a conservation rate structure.

Consumer surplus — The difference between what a commodity is worth to a consumer and what he/she actually pays for it.

Cooling tower — A mechanical device that cools a circulating stream of water by evaporating a portion of it. A cooling tower is part of a system that provides air conditioning or equipment cooling. It usually includes a heat exchanger, recirculating water system, fans, drains, and make-up water supply.

Cooling tower makeup — Water added to the recirculating cooling tower water stream to compensate for water evaporation losses.

Cooling water blowdown — Procedure used to reduce total dissolved solids by removing a portion of poor-quality recirculating water.

Cooling water evaporation — Cooling water recycling approach in which water loses heat when a portion of it is evaporated.

Cost-effectiveness — A comparison of total benefits against total costs.

Costs — The resources needed for a course of action.

Crop coefficient (Kc) — A factor used to adjust reference evapotranspiration and calculate water requirements for a given plant species (also called *plant factor* or *landscape coefficient*).

Cubic feet per second (cfs) — A rate of flow; the volume, in cubic feet, of water passing a reference point in one second.

Cubic foot — A measurement of water equal to 7.48 gallons.

Customer class — A group of customers (residential, commercial, industrial, wholesale, etc.) defined by similar characteristics or patterns of water usage.

Customer surveys — A measure to determine the implementation rate of modifications recommended by a specific measure or program that the customer participated in, and to gauge the participant's satisfaction with the program. A questionnaire sent as a post card. Survey questions could include the extent to which modifications were implemented, the customer's satisfaction with new equipment, satisfaction with utility personnel and/or utility-appointed contractors who performed the evaluation and/or processed the rebate and the customer's perceived change in their water bill.

Declining block rate — A water pricing structure in which customers are charged less per unit of water as consumption increases. This pricing structure discourages conservation by rewarding high water users.

Declining (decreasing) block rate structure — A pricing structure in which the amount charged per unit of water (e.g., dollars per 1,000 gallons) decreases as customer water consumption increases. This type of rate structure is not considered to be water conserving,

Decreasing block rate — Pricing that reflects per-unit costs of production and delivery that go down as customers consume more water.

Dedicated metering — Metering of water service based on a single type of use, such as metering for landscape irrigation separately from interior domestic use.

Deep percolation — The movement of water by gravity downward through the soil profile beyond the root zone.

Demand management — Measures, practices, or incentives deployed by utilities to change the pattern of demand for a service by its customers or slow the rate of growth for that service.

Demand-side measures — In the water industry, programs that encourage customers to modify the amount or timing of water use. These measures may include encouraging customers to implement hardware or behavior changes, or change the volume or timing of their use, depending on the time of day or time of year.

Distribution uniformity (DU) — In irrigation is a measure of how uniformly water is applied to the area being watered, expressed as a percentage. Often calculated when performing an irrigation audit and widely used as a measure of irrigation zone efficiency.

Drip irrigation — A type of micro-irrigation systems that delivers water is slow drips to plants through a network of plastic pipes and emitters.

Drought — Climatic condition in which there is insufficient soil moisture available for normal vegetative growth for an extended period of time.

Drought rate structure — An element of a utility rate structure intended to provide an economic incentive to reduce water use during times of drought.

Dual and multiple programming — The capacity of an irrigation controller to schedule the frequency and duration of irrigation cycles to meet varying water requirements of plants served by a system. Grouping plants and laying out irrigation stations by similar water requirements facilitates multiple programming.

Dual-flush toilet — A toilet designed to use a lower volume of water to flush liquid wastes and a higher volume of water to flush solid wastes.

DWSRF — Drinking water State Revolving Fund, a USEPA program.

Dye test — A test for water leaks, specifically by putting dye in a toilet tank to see if it appears in the bowl.

Early closure flapper — A toilet flapper valve that closes sooner than normal to reduce the volume of water flushed.

Effective precipitation (EP) — The portion of total rainfall that is available for use by the plant.

Efficiency — The use of a resource that maximizes the benefit and minimizes consumption of the resource.

Efficiency standard — A value or criteria that establishes target levels of water use for a particular activity.

Effluent — Wastewater, treated or untreated, that flows out of a treatment plant, sewer, or industrial outfall.

Emitter — A drip irrigation component that dispenses water to plants at a slow and predictable rate, measured in gallons per hour.

End use — A fixture, appliance, or other specific object or activity that uses water.

Energy Policy Act (1992) (EPAct) — A federal law enacted by President George Bush that established maximum allowable water-use requirements for toilets, urinals, showerheads, and faucets manufactured and sold in the United States.

Established landscape — A landscape that has been in place for an extended period of time where the roots of the plants are well developed.

Estimated Water Use (EWU) — The amount of water estimated needed by the landscape during one year.

Estuary — The lower course of a river where its flow is commingled by the sea, resulting in brackish water.

ET factor — A factor used to set a landscape water efficiency goal. Also know as an *adjustment factor.*

ETO — The evapotranspiration of a broad expanse of adequately watered cool-season grass 4 to 6 inches in height. A standard measurement for determining maximum water allowances for plants so that regional differences in climate can be accommodated.

Evaporation — The process by which water changes from liquid to vapor.

Evapotranspiration (ET) rate — A measure of the amount of water required to maximize plant growth. This measure is calculated from climatic conditions and factors such as temperature, solar radiation, humidity, wind, time of year, precipitation, etc.

External costs and benefits — An external cost is when one party adversely affects another party either by reducing its productivity or well being. An external benefit is where one party beneficially affects another party either by increasing its productivity or its well being, or lowering its costs.

Externalities — External costs and benefits.

Faucet aerator — A flow reduction device that screws on the end of the kitchen or lavatory faucet to add air to the water flow.

Faucet restrictor — A device inserted into a faucet that forces water through a smaller orifice for the purpose of reducing the flow rate.

Fixed charge — The portion of a water bill that does not vary with water use.

Fixed costs — Operation and maintenance costs (such as energy) that are independent of the amount of water used. Fixed costs are not affected by conservation activities, while variable costs are influenced by conservation. Costs for a utility that do not vary with the amount of water produced, delivered, and sold to customers.

Flapper valve — A pliable valve in the opening at the bottom of a toilet tank that regulates water flow into the toilet bowl.

Flat rate structure — A fee structure in which the price of water or reclaimed water per unit is constant, regardless of consumption. This type of rate structure is not considered to be water conserving.

Flow rate — The rate at which a volume of water flows through pipes, valves, etc. in a given period of time. Often reported as cubic feet per second (cfs) or gallons-per-minute (gpm).

Flow restrictor — A washer-like disk that fits inside a faucet or showerhead and reduces the water flow rate.

Flushometer toilet — A tankless toilet with the flush valve attached to a pressurized water supply pipe. These toilets are typically found in large institutional and commercial buildings such as schools, airports, office buildings, etc.

Free riders — Customers who would have taken action on their own within a year of receiving a utility-provided financial incentive.

gpc — Gallons per cycles.

gpcd — Gallons per capita per day.

gpd — Gallons per day.

gpf — Gallons per flush.

gph — Gallons per hour.

gphd — Gallons per household per day.

gpl — Gallons per load (of laundry or dishes).

gpm — Gallons per minute.

gpy — Gallons per year.

Gravity-flush toilet — The standard tank style of toilet that uses water (at standard gravitational pressure) to perform flushing functions.

Graywater — Domestic wastewater composed of wash water from kitchen sinks, bathroom sinks and tubs, clothes washers, and laundry tubs that can be used for non-potable purposes such as irrigation.

Green building — Sustainable buildings that are the outcome of designs that focus on increasing the efficiency of resource use — energy, water, and materials — while reducing building impacts on human health and the environment during the building's life cycle, through better siting and design.

Green industry — The trades, professions, and disciplines related to landscape and irrigation research, design, installation, and management.

Groundwater — Water that has seeped beneath the earth's surface and is stored in the pores and spaces between alluvial materials (sand, gravel, or clay).

Growing season — The period, often the frost-free period, during which the climate is such that crops can be produced.

H-axis clothes washer — Horizontal-axis (front-loading) clothes washer.

High efficiency clothes washer (HECW) — A type of clothes washer meeting certain water and energy standards. They often involve a design where the tub axis is more nearly horizontal than vertical. Clothes are tumbled through water that only fills a fraction of the tub. Also known as a *horizontal axis, tumble action*, or *front-loading clothes washer.*

High water-use landscape — A landscape made up of plants, turf and features that requires 50 to 80 percent of the reference evapotranspiration to maintain optimal appearance.

High water-using plants — Plants with a crop coefficient greater than 0.7.

Hot water on demand system — A system of pumping hot water more quickly from the water heater to the fixture calling for water for the purpose of reducing the wait time (and associated waste) for hot water.

Hydrozone — A portion of a landscaped area comprising plants with similar water requirements.

ICI — Institutional, Commercial, Industrial customer sector. Also referred to as *CII.*

Inclining block rate — A commodity rate whose unit price increases with increasing water use.

Increasing block rate — A water pricing structure in which customers are charged more per unit of water as consumption increases. This type of pricing structure encourages conservation by penalizing high water users.

Individual metering — The installation of meters for each individual dwelling unit as well as separate common area metering with the local water utility providing customer read, bill and collect services.

Industrial user — Water users that are primarily manufacturers or processors of materials as defined by the Standard Industrial Classifications (SIC). Code numbers 2000 through 3999.

Infiltration rate — The rate of water entry into the soil expressed as a depth of water per unit of time in inches per hour or feet per day. The infiltration rate changes with time during irrigation.

Inflation — The rate of change in a price index.

Informative billing — System of providing water utility customers with useful information on the relationship between the amount of water they use and the cost associated with that use. Examples of the information include the utility's rate structure, amount of water used in the current month, amount of water used in the previous month, amount of water used in the same month of the previous year, information on the average usage of all customers in the same customer class, seasonal rates and applicable months, drought rates, information on conserving water, or other information deemed appropriate by the utility.

Institutional user — Water-using establishment dedicated to public service. This includes schools, churches, hospitals, and government facilities. All facilities serving these functions are considered institutional regardless of ownership.

Integrated resource planning — A planning process emphasizing least-cost principles and balanced consideration of supply and demand management options for meeting water needs.

Invasive plant — A nonindigenous plant that invades and takes over substantial areas of an ecosystem.

Inverted block rates — See inclining (increasing) block rate.

Irrigated area — The portion of a landscape that requires supplemental irrigation, usually expressed in square feet or acres.

Irrigation audit — An on-site evaluation of an irrigation system to assess its water-use efficiency as measured by distribution uniformity, irrigation schedule, and other factors.

Irrigation controller — A mechanical or electronic clock that can be programmed to operate remote-control valves to control watering times.

Irrigation cycle — A scheduled application of water by an irrigation station defined by a start time and its duration. Multiple cycles can be scheduled, separated by time intervals, to allow infiltration of applied water.

Irrigation efficiency — The ratio of the average depth of irrigation water that is beneficially used to the average depth of irrigation water applied, expressed as a percent. Beneficial uses include satisfying the soil water deficit and any leaching requirement to remove salts from the root zone.

Irrigation only accounts — Accounts with a separate meter dedicated to non-sewered uses such as landscape irrigation or cooling towers.

Irrigation requirement — Quantity of water, exclusive of effective precipitation, that is required for maintaining a landscape.

Irrigation scheduling — The process of developing a schedule for an automatic irrigation system that applies the right amount of water, matched to the plant needs, which varies daily, weekly, or seasonally.

Irrigation timer — A device that can be programmed to regulate the time and duration of irrigation; a sprinkler clock.

Landscape irrigation auditor — A person who has had landscape water audit training and passed a certification exam.

Landscape water budget (LWB) — A volume of applied irrigation water expressed as a monthly or yearly amount, based on ETO and the plant material being watered.

Landscape water requirement — A measure of the supplemental water required to maintain the optimum health and appearance of landscape plants and features.

Leak detection — Systematic methods for identifying water leakage from pipes, plumbing fixtures, and fittings.

Lifeline rate — A minimum, sometimes subsidized water rate created to help meet basic human needs.

Limited turf areas — Restriction of turfgrass to a prescribed fraction of the landscape area.

Long-term reductions — Conservation measures undertaken to reduce water use over the long term (more than one year). Long-term reductions are the focus of this guidebook.

Low-flow faucet — A faucet fixture that meets 1992 EPAct standards (2.2 gpm or less at 80 psi).

Low-flow showerhead — A showerhead that meets 1992 EPAct standards (2.5 gpm or less at 80 psi).

Low-flow toilet — A 3.5 gpf toilet, as mandated by California in a 1977 law that took effect 1980.

Low-flow faucet — A faucet that uses no more than 2.5 gpm at 80 psi.

Low-flow showerhead — A showerhead that requires 2.5 gpm or less.

Low-flush toilet — A toilet that requires 1.6 gal of water per flush or less.

Low-volume irrigation — The use of equipment and devices specifically designed to allow the volume of water delivered to be limited to a level consistent with the water requirements of the plant being irrigated and to allow that water to be placed with a high degree of efficiency.

Low-volume urinal — A urinal that uses no more than 1.0 gallons per flush.

Low water-use landscape — Use of plants that are appropriate to an area's climate and growing conditions.

Low water-use plants — Plants that require less than 30 percent of reference ET to maintain optimum health and appearance.

Marginal cost — The additional cost incurred by supplying one more unit of water.

Marginal-cost pricing — A rate design method where prices reflect the costs associated with producing the next increment of supply.

Market penetration — The degree of actual usage that a conservation measure achieves among the intended user group or market.

Master meter — A single meter that measures utility usage for an entire property, or an entire building, which usually includes common areas.

Measure — An action, behavioral change, device, technology, or improved design or process implemented to reduce water loss, waste, or use. It should be noted that the value and cost-effectiveness of a water efficiency measure must be evaluated in relation to its effects on the use and cost of other natural resources (e.g. energy and/or chemicals).

Medium water-use plants — Plants that require 30 to 50 percent of reference ET to maintain optimum health and appearance; with a crop coefficient of 0.4 to 0.6.

Meter (water) — An instrument for measuring and recording water volume.

mgd — Million gallons per day.

mgy — Million gallons per year.

Microirrigation — An irrigation system with small, closely spaced outlets used to apply small amounts of water at low pressure.

MOU — Memorandum of understanding.

Mulch — A protective covering of various substances, usually organic, such as wood chips, placed on the soil surface around plants to reduce weed growth and evaporation and to maintain even temperatures around plant roots.

Multi-family (MF) — Residential housing with multiple dwelling units, such as apartments and condominiums.

Multiple start times — An irrigation controller's capacity to accept programming of more than one irrigation start-time per station per day.

NAICS (formally SIC codes) — North American Industry Classification System. A consolidation of the codes for the US, Canada and Mexico. Produced by the US Office of Management and Budget.

Native plants — Plants that are indigenous to a region and require little or no supplemental irrigation after establishment.

Nonpotable irrigation source replacement or rebates — A water conservation quantifiable BMP that provides reclaimed water or rebates (for use of other nonpotable water sources for irrigation) to the single-family and multi-family residential sectors and/or the nonresidential sector. Categories include: (1) groundwater from wells; and (2) other irrigation sources; does not include reclaimed water.

Nonrevenue water — In a distribution system water audit, nonrevenue water equals the volume of unbilled authorized consumption (water for fire fighting, system flushing, and similar uses) added to real losses and apparent losses.

Overspray — Application of water via sprinkler irrigation to areas other than the intended area.

Peak demand — The maximum amount of water required to meet demand during a specified time period, usually expressed in terms of monthly, daily, or hourly peak demand.

Peak use — The maximum demand occurring in a given period, such as hourly or daily or annually.

Peak/off-peak rates — Rates charged in accordance with the most and least popular hours of water use during the day.

Penetration rates — The extent to which a water efficiency measure is actually implemented.

Per capita residential use — Average daily water use (sales) to residential customers divided by population served.

Per capita use — The amount of water used by one person during one 24-hour period. Typically expressed as gallons per capita per day (gpcd).

Potable water — Water that meets federal and state water quality standards for water delivered to utility customers.

Pounds per square inch (psi) — A standard measure of water pressure.

Precipitation rate — Application rate for sprinkler irrigation, generally measured in inches-per-hour.

Pressure reducer — A water system component that reduces the downstream pressure of water, often used in irrigation systems, always used in drip systems.

Pressure regulating valve — 1) A device, often installed downstream of the customer meter, to reduce high pressures to a set amount. Often required where the existing system pressure exceeds 85 psi. 2) A device installed on input water supply mains or irrigation systems to regulate water pressure in a zone or district metered area (DMA) to protect against pressure surges and to control leakage.

Pressure regulator — A device used to limit water pressure.

Price elasticity of demand — A measure of the responsiveness of customer water use to changes in the price of water; measured by the percentage change in price.

Public service announcement (PSA) — An inexpensive or free advertisement or message on mass media that serves the public good.

Rain sensor — A device that automatically shuts off an irrigation system after a set amount of precipitation falls.

Rain shutoff device — A device connected to an irrigation controller that overrides scheduled irrigation when significant precipitation is detected.

Rain switch — A simple on/off switch on an irrigation system that facilitates the shutdown of the system during a rainstorm.

Rainwater harvesting — The capture and use of runoff from rainfall.

Rationing — Mandatory water use restrictions typically imposed during a drought.

Recirculating cooling water — Recycling cooling water to greatly reduce water use by using the same water to perform several cooling operations.

Reclaimed water — Municipal wastewater effluent that is given additional treatment and distributed for reuse in certain applications. Also referred to as *recycled water.*

Recycled water — A type of reuse water usually run repeatedly through a closed system; sometimes used to describe reclaimed water.

Reference evapotranspiration (ETO) — The water requirements of a standardized landscape plot, specifically, the estimate of the evapotranspiration of a broad expanse of well-watered, 4-to-7 inch-tall cool-season grass.

Residential End Uses of Water Study (REUWS) — The Residential End Uses of Water Study published by the American Water Works Association Research Foundation in 1999.

Residential water use — Water use in homes and apartments.

Retention rate — The percent of devices that remain in-place over time after initially being installed or distributed.

Retrofit — 1) Replacement of existing water using fixtures or appliances with new and more efficient ones. 2) Replacement of parts for a fixture or appliance to make the device more efficient.

Retrofit kit away — A water conservation measure or quantifiable BMP (depending on program implementation) that is administered as a customer-installation program or a door-to-door installation program. Retrofit kits typically contain low flow showerhead(s), faucet aerators, toilet leak-detection dye tablets, and informational brochures on how to identify, measure, and/or fix leaks.

Retrofit on resale — A regulation that requires plumbing fixtures to be upgraded to current code at the time property is sold.

Return flow — That portion of the water diverted from a stream that finds its way back to the stream channel, either as surface or underground flow.

Reuse — Use of treated municipal wastewater effluent for specific, direct, beneficial uses. See *reclaimed water.* Also used to describe water that is captured on-site and utilized in a new application.

Revenue-producing water — Water metered and sold.

Rotors — Irrigation rotors are oscillating sprinklers that are designed to water larger areas than standard sprinklers.

Runoff — The portion of precipitation, snow melt, or irrigation that flows over the soil, eventually making its way to surface water supplies.

Seasonal water rates — A rate that varies depending on the time of the year. Seasonal rates can be used in conjunction with any rate structure, including flat rates and uniform, decreasing, or increasing block rates. A water pricing structure that discourages consumption during peak use periods by imposing significantly higher rates during these periods for consumption above baseline water use.

Self-closing faucet — A faucet that automatically shuts off the water flow after a designated amount of time, usually a few seconds.

Service area (territory) — The geographic area(s) served by a utility.

Service-connection metering — A water conservation measure where the water supplier meters at each service connection.

Short-term reductions — Conservation measures undertaken to reduce water use over the short term (i.e., in response to drought or emergency conditions). Short-term emergency measures are not addressed in this guidebook.

SIC Code (Standard Industrial Classifi-cation) — A system devised by the federal government to classify industries by their major type of economic activity. The code may extend from two to eight digits. This term has been superseded by the NAICS.

Simple payback period — The length of time over which the cost savings associated with a conservation measure must accrue to equal the cost of implementing the measure.

Simple water budget — A water budget that is the product of reference evapotranspiration, irrigated area, and a conversion factor.

Single-family (SF) unit — A residential dwelling unit built with the intent of being occupied by one family. It may be detached or attached (e.g., townhouses).

Soil amendment — Organic and inorganic materials added to soils to improve their texture, nutrients, moisture holding capacity, and infiltration rates.

Soil improvement — The addition of soil amendments.

Soil moisture deficit — The amount of water required to saturate the plant root zone at the time of irrigation, expressed as a depth of water in inches or feet.

Soil moisture replacement — The amount of water applied to replace a portion of all the soil moisture deficit, expressed as a depth of water in inches or feet.

Soil moisture sensor — A device placed in the ground at the plant root zone depth to measure the amount of water in the soil. Soil moisture sensors are also used to control irrigation and signal whether watering is required or not.

Spray head — A sprinkler irrigation nozzle installed on a riser that delivers water in a fixed pattern. Flow rates of spray heads are high relative to the area covered by the spray pattern.

Sprinkler heads — Devices that distribute water over a given area for irrigation (or to put out fires). The primary purpose of sprinklers, however, is to get golfers wet on cold mornings.

Sprinkler irrigation — Overhead delivery of water spray heads, stream rotors, or impact heads. Precipitation rates will vary depending on system layout and type of head used.

Sprinkler run time — The minutes of irrigation per day, based on the weekly irrigation requirement and irrigation days per week.

Sprinkler station — A group of sprinklers controlled by the same valve.

Sprinkler valve — The on-off valve, usually electric, that controls an irrigation or sprinkler station.

Stream rotors — Sprinkler irrigation heads that deliver rotating streams of water in full or partial circles. Some types use a gear mechanism and water pressure to generate a single stream or multiple streams. Stream rotors have relatively low precipitation rates, and multiple stream rotors can provide matched precipitation for varying arc patterns.

Structured plumbing system — Properly sized and well insulated hot water main and hot water risers, including a dedicated hot water main segment connecting the farthest hot water point of use to the water heater.

Subirrigation — Applying irrigation water below the ground surface either by raising the water table within or near the root zone, or by use of a buried perforated or porous pipe system that discharges directly into the root zone.

Submetering — The practice of using meters to measure master-metered utility consumption by individual users. Also, see partial-capture submetering and total-capture submetering.

Subsidence — The lowering of ground surface due to extraction of material from subsurface. Can be caused by water or oil extraction from the ground.

Subsurface drip irrigation — The application of water via buried pipe and emitters, with flow rates measured in gph.

Subsurface irrigation — Applying irrigation water below the ground surface either by raising the water table within or near the root zone, or by use of a buried perforated or porous pipe system that discharges directly into the root zone.

Sunken costs — Costs that have already been incurred and are not reversible.

Supplemental irrigation — The application of water to a landscape to supplement natural phenomena.

Supply-side measures — Increasing water supply by developing more raw water, generally building reservoirs and canals or drilling groundwater wells.

Surcharge — A special charge included on a water bill to recover costs associated with a particular activity, facility, use, or to convey a message about water prices to customers.

System audit — A water conservation measure where the water supplier performs a systematic accounting of water throughout the production, transmission, and distribution facilities of a water supply system.

System Safe Yield — The amount of water that can be withdrawn annually without depleting the water source. Safe yield is usually defined by hydrologic studies conducted at the time the adequacy of the source is evaluated.

Tiered pricing — Increasing block-rate pricing.

Time-of-day pricing — Pricing that charges users relatively higher prices during utilities' peak use periods.

Toilet displacement device — A toilet retrofit device (such as a dam, bag, bottle, or rock) used to displace water in the toilet tank to reduce the volume required for flushing.

Toilet flapper — A pliable valve in the opening at the bottom of a toilet tank that regulates water flow into the toilet bowl.

Toilet tank fill cycle regulator — A device that reduces the amount of water that goes into the overflow tube and hence into the toilet bowl during a toilet flush.

Transpiration — The passing of water through living plant membranes into the atmosphere.

Turfgrass — Hybridized grasses that, when regularly mowed, form a dense growth of leaf blades and roots.

Ultra Low Flush Toilet (ULFT) — A toilet that flushes with 1.6 gal or less.

Unaccounted-for water — The difference between the amount of water entering the distribution system and the amount of water supplied to customers. Sources of unaccounted-for water include authorized uses (such as fire hydrants and main flushing) and unauthorized uses (illegal connections). The remaining unaccounted-for water results from meters out of calibration or from leaks in the distribution system.

Under-irrigation — The difference between the water stored in a plant root zone during irrigation and the amount needed to refill the root zone to field capacity.

Uniform block rate — A commodity rate that does not vary with the amount of water use.

Uniformity — See *distribution uniformity.*

United States Bureau of Reclamation (USBR, the Bureau) — Federal agency that built and operates water projects in the western United States. Part of the Department of Interior.

Unmetered supply — A water supply system that does not use meters to measure water usage by individual service connections.

User class — Customers having similar characteristics (commercial, single-family residential, etc) grouped together for billing or program purposes.

Utility — Public water service provider.

Valve — A device used to control the flow of water. Isolation valves are used to shut-off water for repairs. Control valves turn on and off the water to the individual circuits of sprinklers or drip emitters. Check valves allow the water to flow in only one direction. Master valves are located at the water source and turn on and off the water for the entire irrigation system when not in use.

Valve zone — An area where irrigation is all controlled by a single control valve. Each valve zone must be within only one hydrozone.

Variable charge — The portion of a water bill that varies with water use; also known as a *commodity charge.*

Variable costs — Operation and maintenance costs (such as energy and chemicals) that are dependent on the amount of water used. Variable costs are affected by conservation activities, while fixed costs are not.

Volume-based rates — Rates for water that are based on the amount of water used; may or may not be a water-conserving rate structure.

Water audit — An analysis of customer water use practices. A water audit typically involves identifying water uses, discussing water use practices with the customer, and providing information and assistance with water-saving measures.

Water banking — A process whereby unused water allocations are held in storage and made available for future water allocations.

Water bill inserts — A water conservation measure where the water supplier includes inserts in their customers' water bills that provide information on water use and costs and/or provide tips for home water conservation.

Water budget approach — A method of establishing water-efficiency standards for landscapes by providing the water necessary to meet the ET of the landscaped area.

Water conservation — Any beneficial reduction in water use or in water losses. Activities designed to reduce the demand for water, improve efficiency in use, and reduce losses and waste of water.

Water conservation incentive — An effort designed to promote customer awareness about reducing water use and motivate customers to adopt specific conservation measures.

Water conservation measure — An action, behavioral change, device, technology, or improved design or process implemented to reduce water loss, waste, or use.

Water demand — Water requirements for a particular purpose, as for irrigation, drinking, toilet flushing, bathing, clothes washing, etc.

Water efficiency — Accomplishment of a function, task, process, or result with the minimal amount of water feasible; an indicator of the relationship between the amount of water required for a particular purpose and the quantity of water used or delivered.

Water efficiency measure — A specific tool or practice that results in more efficient water use and thus reduces water demand.

Water efficiency standard — Criterion creating maximum or acceptable levels of water use.

Water-efficient landscape — A landscape that minimizes water demand through design, installation, and management.

Water feature — A pool, fountain, water sculpture, waterfall, or other decorative element that includes water. Many water features recycle water thus reducing consumption.

Water harvesting — The capture and use of runoff from rainfall.

Water holding capacity — Amount of soil water available to plants.

Water losses — Metered source water less metered and authorized unmetered water use.

Water meter size — Normally corresponds to the pipe bore, for example 1 inch. For some models a second designation refers to the matching pipe end connections.

Water recycling — The treatment of urban wastewater to make it reusable for a specific beneficial purpose.

Water Research Foundation — A nonprofit organization that sponsors research for the drinking water supply profession. Formerly known as Awwa Research Foundation (AwwaRF).

Water right — A legal entitlement authorizing water to be diverted from a specified source and put to beneficial, nonwasteful use. It is a property right, but the holder does not possess the water itself – they possess the right to use it. The primary types of water rights are appropriative and riparian.

There are also prescriptive (openly taking water to which someone else has the right) and pueblo (a municipal right based on Spanish and Mexican law).

Water surcharge — Imposition of a higher rate on excessive water use.

Water use efficiency — A measure of the amount of water used versus the minimum amount required to perform a specific task. In irrigation, the amount of water beneficially applied divided by the total water applied.

Water-use profile — A quantitative description (often displayed graphically) of the different water uses at a residence, business site, or utility service area.

Water Waste Prohibition — A water conservation measure where the water supplier develops and enforces an ordinance prohibiting gutter flooding, single-pass cooling systems in new connections, nonrecirculating systems in all new conveyer car wash and commercial laundry systems, and nonrecycling decorative water fountains.

Waterless urinal — A urinal that works without water or flush valves. Instead a cartridge filled with a sealant liquid is placed in the drain.

Water-use evaluation — A systematic accounting of water uses by end users (e.g., residential, landscape, commercial, industrial, institutional, or agricultural customers), usually conducted to identify potential opportunities for water use reduction through efficiency measures or improvements; also referred to as *end-use audit* or *water-use audit*.

WaterSense — The US Environmental Protection Agency's certification and labeling system for water efficient products www.epa.gov/WaterSense

Watershed — A land area, defined by topography, soil, and drainage characteristics, within which raw waters are contained. They can collect to form a stream or percolate into the ground.

Weather station — A facility where meteorological data are gathered.

Xeriscape™ — A trademarked term denoting landscaping that involves the selection, placement, and care of low-water-use plants. Xeriscape is based on seven principles: proper planning and design, soil analysis and improvement, practical turf areas, appropriate plant selection, efficient irrigation, mulching, and appropriate maintenance.

References

Abbey, B. 2009. Green Law Web Site Louisiana State University http://www.greenlaws.lsu.edu/

Alliance for Water Efficiency (AWE) Clearinghouse. 2009. www.allianceforwaterefficiency.org/. Last accessed July 8, 2009.

American Public Works Association. 1981. *Planning and Evaluating Water Conservation Measures*. Special Report No. 48. Chicago, Ill.

American Water Works Association (AWWA). 2000. M1 *Principles of Water Rates, Fees, and Charges*. Denver, Colo: AWWA.

AWWA. 2009. M24 *Planning for the Distribution of Reclaimed Water*. Denver, Colo.: AWWA.

AWWA. 2009. M36 *Water Audits and Leak Detection*. Denver, Colo.: AWWA.

AWWA. 2006. M52 *Water Conservation Programs—A Planning Manual*. Denver, Colo.: AWWA.

American Water Works Association, Pacific Northwest Section. 1993. *Water Conservation Guidebook for Small and Medium-Sized Utilities*. Denver, Colo.: AWWA.

Anderson, K.M. 2004. An Investigation into What Planning Departments and Water Authorities Can Learn form Eleven Communities' Waterwise Landscaping Ordinances. *Masters Thesis, Community and Regional Planning*, University of Oregon. Eugene, Ore.

Aqua Conserve. 2003. Residential Landscape Irrigation. *Study using Aqua ET Controllers*. http://www.cuwcc.org/et_controllers.

Aquacraft. 2004. National Multiple Family Submetering and Allocation Billing Program Study. http://www.aquacraft.com/Download_Reports/Submetering_and_Allocation_Billing_Final_Report.pdf.

Arizona Municipal Water Users Association. 2008. Facility Manager's Guide to Water Management. http://www.amwua.org/conservation/facility_managers_guide.pdf

Ash, T. 2002. Using ET Controller Technology to Reduce Demand and Urban Water Run-Off: Summary of the Technology, Water Savings Potential & Agency Programs.

American Water Works Association Water Sources Conference Proceedings. Denver, Colo.: AWWA.

AWE Water budget rate structures.pdf. 2008.

Beecher, J.A. 2002. *Survey of State Agency Reporting Practices: Final Report to the American Water Works Association. January 2002*. Denver, Colo.: AWWA.

Bennett, R.E. and M.S. Hazinski. 1993. *Water-Efficient Landscape Guidelines*. Denver, Colo.: AWWA.

Bracciano, D., R.Pensa, J. Walkinshaw. 2004. What's the Flap over Flappers? Using Research, Marketing, and Implementation Strategies to Secure ULF Toilet Savings. *Proceedings of the AWWA Water Sources Conference*. Denver, Colo.: AWWA.

British Columbia Ministry of Environment. 2009. Living Water Smart: British Columbia's Water Plan. www.livingwatersmart.ca. Last accessed May 24, 2009.

Broustis, D. 2006. Maximizing Residential Clothes Washer Water Savings. *AWWA Sources Conference*. Denver, Colo.: AWWA.

Brown, C. 2000. Water Conservation in the Professional Car Wash Industry. *A Report for the International Carwash Association*. Chicago, Ill.

California Department of Water Resources. 2008. Examples of Complete Urban Water Management Plans. http://www.owue.water.ca.gov/urbanplan/uwmp/uwmp.cfm. Last accessed September 16, 2008.

California Urban Water Conservation Council. 2005. BMP Costs & Savings Study: A Guide to Data and Methods for Cost-Effectiveness Analysis of Urban Water Conservation Best Management Practices. A & N Technical Services, California Urban Water Council Website. www.cuwcc.org. Last accessed July 8, 2009.

Canadian Mortgage and Housing Corporation (Forthcoming). 2001. *Guide for the Development of Municipal Water Efficiency Plans in Canada*. Golden Valley, Minn.: Veritec.

Cardenas-Lailhacar, B., & Dukes, M. D. 2008. Expanding disk rain sensor performance and potential irrigation savings. *Journal of Irrigation and Drainage Engineering*, 134(1): 67-73.

Chesnutt, T., Fiske, G. , Beecher, J. and D. Pekelney. 2007. *Water Conservation Programs for Integrated Water Management*. Denver, Colo.: AWWARF. .

Chesnutt, T.W., A. Bamezai, C.N. McSpadden. 1992. The Conserving Effect of Ultra-Low Flush Toilet Rebate Programs. Santa Monica, Calif.: A&N Technical Services, Inc.

Chesnutt, T. 1997. *Designing, Evaluating, and Implementing Conservation Rate Structures*. Sacramento, Calif.: CUWCC.

Chinery, G. 2006. USEPA ENERGY STAR for Homes. Policy recommendations for the HERS Community to consider regarding HERS scoring credit due to enhanced effective energy factors of water heaters resulting from volumetric hot water savings due to conservation devices/strategies. Washington, D.C.: USEPA. www.energystar.gov.

City of Santa Cruz Water Conservation Plan. 2005. http://www.ci.santa-cruz.ca.us/wt/conservation/

Colorado Water Conservation Board (CWCB). 2005. *Water Conservation Plan Development Guidance Document*. http://cwcb.state.co.us/Conservation/RelatedInformation/Publications/WaterConservationPlanDevelopmentGuidanceDocument/

Community Based Social Marketing Web Site. 2009. http://www.cbsm.com. Last accessed September 16, 2008.

Cooley, H., T. Hutchins-Cabibi, M. Cohen, P. H. Gleick, and M. Heberger. 2007. Hidden Oasis: Water Conservation and Efficiency in Las Vegas. Oakland, Calif.: Pacific Institute www.pacinst.org/reports/las_vegas.

Davis, N.H. 2009. Personal Communication. Tampa, Fla.: Hillsborough County Water Resources.

DeOreo, W., A. Dietemann, T. Skeel, P. Mayer, D. Lewis, and J. Smith. 2001. Retrofit Realities. *Journ. AWWA*, 93(3):58-72.

Dickinson, M. and E. Gardener. 2004. Conserving Water in Landscapes: Potentially Your Greatest Savings." Landscape Workshop. AWWA Annual Conference. Denver, Colo.: AWWA.

Dukes, M. D., B. Cardenas-Lailhacar, B., and G.L. Miller. 2005. *Irrigation Research at UF/IFAS*. Retrieved June 27, 2008, from Institute of Food and Agricultural Sciences: http://irrigation.ifas.ufl.edu/SMS/pubs/June05_Resource_sensor_irrig.pdf.

East Bay Municipal Utility District. 2009. *WaterSmart Guidebook: A Water Use Efficiency Plan and Review Guide*. Oakland, Calif. http://www.cuwcc.org/uploadedFiles/

Environmental Protection Agency WaterSense Program (USEPA). 2009. Draft WaterSense High-Efficiency Flushing Urinal Specification Supporting Statement. epa.gov/watersense/docs/urinal_suppstat508.pdf,

USEPA. 1992. *Statement of Principles on Efficient Water Use*. Washington D.C.

USEPA. 2009. *WaterSense* website www.epa.gov/watersense/

Federal Energy Management Program (FEMP). 2009. www.eere.energy.gov/femp/ last accessed July 8, 2009.

FEMP. 2009. Water Efficiency BMP #9 Single-Pass Cooling Equipment http://www1.eere.energy.gov/femp/water/water_bmp9.html,

Gardener E. 2009. Personal Communication. Denver Water, Denver, Colo.

Gauley, W. 2002. *Water Efficiency Plan*. City of Toronto Works and Emergency Services. http://www.toronto.ca/watereff/plan.htm

Gauley, W. and J. Koeller. 2004. Replacement Toilet Flappers... are they flushing at 1.6 gallons? *Proceedings* from the American Water Works Association Water Sources Conference. Denver, Colo.: AWWA.

Gauley, W. and J. Koeller. 2003. Unified North American Requirements for Toilet Fixtures (UNAR). http://www.awwa.org/waterwiser/watch/UNARdescription1.pdf

Gauley, W. 2008. *Myths in Field of Water Efficiency & Some Innovative Products*. AUMA Municipal Water Conservation Workshop. Red Deer, Alb., Can. http://www.auma.ca/live/digitalAssets/23/23208_Myths_in_Field_of_Water_Efficiency_Bill.pdf

Gelt, J. 2009. *Home Use of Graywater, Rainwater Conserves Water--and May Save Money*. Water Resources Research Center. Tucson, Ariz.: University of Arizona. http://ag.arizona.edu/AZWATER/arroyo/071rain.html undated. Accessed April 20, 2009.

Gleick, P. H., D. Haasz, C. Henges-Jeck, V. Srinivasan, G. Wolff, K. K. Cushing and A. MA. . 2003. Waste Not, Want Not: The Potential for Urban Water Conservation in California. *Pacific Institute for Studies in Development, Environment, and Security*. Oakland, Calif.

Green, D. and J. Yingling. 2007. Increase Conservation and Maintain Revenues: Mission (Not) Impossible. *Florida Water Resources Journal*. Publication #FE756.

Green, D. 2008 . July is Smart Irrigation Month… But Not the Month of Highest Water Use, If We Use Rain Sensors. "Using Water Efficiently" Column, *Florida Water Resources Journal* 60(7):74-78).

Green, D. 2009. Knowledge is Power-Informative Billing for Increased Water Use Efficiency. *Florida Water Resources Journal* (61) 11: 42-43.

Gregg, T. 2006. *Austin's Water Conservation Task Force: Outdoor Water Efficiency Programs. Water Sources.* Austin, Texas: City of Austin. www.awwa.org/.../waterwiser

Haman, D. , M. Dukes, and S. Park-Brown "Retrofitting a Traditional In-ground Sprinkler Irrigation System for Microirrigation of Landscape Plants" University of Florida Institute of Food and Agricultural Sciences ABE324. http://edis.ifas.ufl.edu/pdffiles/AE/AE22200.pdf

Hoffman, H.W. 2009. *Alternate On-Site Sources of Water – The New Green Build Buzz.* Proceeding from WaterSmart Innovations Conference. Las Vegas, Nev. www.watersmartinnovations.com.

Hoffman, H.W. 2008. Capturing the water you already have: Using alternate onsite sources. *Journ. AWWA*, 100 (5): 112-116.

Hoffman, H.W. 2007. Commercial Ice Machine Life-cycle Cost Calculator. www.a4we.org/WorkArea/linkit.aspx?LinkIdentifier=id&ItemID=998

Horner, R. 2009. *How Much Water Does a Swimming Pool Use?* Proceeding from WaterSmart Innovations Conference and Exposition. Las Vegas, Nev. www.watersmartinnovations.com.

Irvine Ranch Water District. 2001. Residential Weather-Based Irrigation Scheduling: Evidence from the Irvine "ET Controller" Study. Irvine Ranch Water District, the Municipal Water District of Orange County, and the Metropolitan Water District of Southern California.

Johns, Grace M. 2007. *Financing Water Conservation.* Presentation at AWWA Water Conservation Division Water Conservation Workshop Savannah, Georgia. http://www.awwa.org/waterwiser/references/ConservationWorkshop2007.cfm

Klein, G. 2004. Saving Energy and Water in Residential Hot Water Systems. *AWWA Conference Proceedings.* Denver, Colo.

Koeller, J. 2009. Danger in the Shower - 2008 Forum Looks at Hot Water. *Home Energy Jan/Feb.*

Koeller, J. 2007. Myths about water using products. *The WaterLogue.* California Urban Water Conservation Council, Vol 5, Issue 3.

Koeller, J. 2008. Report on Potential Best Management Practices California Urban Water Conservation Council. Sacramento, Calif. *Efficiency Standards on Water Infrastructure Investments.* Denver, Colo.: American Water Works Association.

Maddaus, L.A., & P.W. Mayer. 2001. *Splash or Sprinkler? Comparing Water Use of Swimming Pools and Irrigated Landscapes.* AWWA Annual Conference Proceedings. Denver, Colo.: American Water Works Association.

Maddaus, W. 1987. *Water Conservation.* Denver, Colo.: American Water Works Association.

Maddaus, W., and M. Maddaus. 2004. *Evaluating Water Conservation Cost-Effectiveness with an End Use Model.* Proc. Water Sources Conference. Austin, Texas: AWWA.

Maddaus, W. 2009. *Evaluating the Benefits and Costs of Proposed Water Conservation.* City of Oceanside,Calif.

Maddaus, M., W. Maddaus, M.l Torre, and R. Harris. 2008. Innovative Water Conservation Supports Sustainable Housing Development. *Journ. AWWA* 100(5):104-111.

Maddaus Water Management. 2009. *Demand Side Management Least Cost Planning Decision Support System or DSS Model*. Santa Barbara, Calif.: Water Efficiency.

Mayer, P., W.B. DeOreo, et al. 2000. *Seattle Home Water Conservation Study: The Impacts of High Efficiency Plumbing Fixture Retrofits in Single-family Homes.* Boulder, Colo.: Aquacraft, Inc.

Mayer, P., W.B. DeOreo, et al. 2003. *Residential Indoor Water Conservation Study: Evaluation of High Efficiency Indoor Plumbing Fixture Retrofits in Single-family Homes in the East Bay Municipal Utility District Service Area*. Boulder, Colo.: Aquacraft, Inc.

Mayer, P., W.B. DeOreo, et al. 2003. Tampa Water Department Residential Indoor Water Conservation Study: THE IMPACTS OF High Efficiency Plumbing Fixture Retrofits In Single Family Homes. Boulder, Colo.: Aquacraft, Inc.

Mayer, P.W., W.B. DeOreo, and P. Lander. 2000. *Seattle Home Water Conservation Study: The Impacts of High Efficiency Plumbing Fixture Retrofits in Single Family Homes*. Boulder, Colo.: Aquacraft, Inc. http://www.aquacraft.com/

Mayer, P.W., W.B. DeOreo, E. Opitz, J. Kiefer, B. Dziegielewski, W. Davis, J.O. Nelson. 1999. *Residential End Uses of Water*. Denver, Colo.: Awwa Research Foundation.

Mayer, P.W., W.B. DeOreo, M.Hayden and R. Davis. 2009. *Evaluation of California Weather-based "Smart" Irrigation Controller Programs.* California Department of Water Resources. Oakland, Calif.: Metropolitan Water District of Southern California and East Bay Municipal Utility District.

McKenzie-Mohr, D. 1999. *Fostering Sustainable Behaviour*. Gabriola Island, B.C.: New Publishers Society.

New Mexico Office of the State Engineer. undated. *Waterwise Guide to Rainwater Harvesting*. http://www.ose.state.nm.us/water-info/conservation/rainwater-harvesting.pdf.

New Mexico Office of the State Engineer. 1999. A Water Conservation Guide for Commercial, Institutional, and Educational Users. www.ose.state.nm.us/water-info/conservation/.../cii-users-guide.pdf

North Carolina Department of Environment and Natural Resources. 2009. *Efficiency Manual for Commercial, Industrial, and Institutional Facilities.* Raleigh, N.C. http://www.p2pays.org/ref/01/00692.pdf.

Pape, T. 2008. Plumbing codes and water efficiency: What's a water utility to do? *Journ. AWWA* 100(5):101-103.

Philpott, B. 2008. *Field Guide to Soil Moisture Sensor Use in Florida. Program for Resources Efficient Communities.* University of Florida. St. Johns River Water Management District. www.sjrwmd.com/floridawaterstar/pdfs/SMS_field_guide.pdf

Planning and Management Consultants, Ltd. 1994. IWR-MAIN Water Demand Analysis Software Version 6.0: User's Manual and System Description. Carbondale, Ill.: Planning and Management Consultants, Ltd.

Planning and Management Consultants, Ltd.; Brown and Caldwell Consultants; Spectrum Economics; Montgomery Watson Consulting Engineers. 1992. *Evaluating Urban Water Conservation Programs: A Procedures Manual.* Report prepared for the California Urban Water Agencies, Sacramento, Calif.

Puckorius, P. 2008. *Water Reuse in Cooling Towers - Current Experiences and Guidelines for Success in Refineries, Power Plants, and HVAC Systems.* Houston, Texas: Cooling Technology Institute TP08-01.

Southern Nevada Water Authority. 2009. website www.snwa.com/

Southwest Florida Water Management District. 2006. *A Guide to Micro-Irrigation for West-Central Florida Landscapes.* www.swfwmd.state.fl.us/publications/files/micro-irrigationguide.pdf .

Southwest Florida Water Management District. 2008. Skip-A-Week Campaign. Brooksville, Fla. http://www.swfwmd.state.fl.us/news/

Sovocool, K. A. and J. L. Rosales. 2001. *A Five-Year Investigation into the Potential Water and Monetary Savings of Residential Xeriscape in the Mojave Desert.* AWWA Annual Conference Proceedings. Denver, Colo. http://www.snwa.com/assets/pdf/xeri_study_preliminary.pdf

Sovocool, K. 2005. *Xeriscape Conversion Study: Final Report.* Las Vegas, Nev.: Southern Nevada Water Authority,

Sunnyslope County Water District. 2009. Hollister, Calif.: http://www.sscwd.org/ordinance.html.

Tampa Bay Water. 2005. Evaluating Implementation of Multiple Irrigation and Landscape Ordinances in the Tampa Bay region. Tampa Bay, Fla.: Hazen and Sawyer. http://www.tampabaywater.org/documents/conservation/41040-TBW-Eval%20of%20Irrig%20Report%205-2-05.pdf

Thornton, J., R. Sturm, and G. Kunkle. 2008. *Water Loss Control.* 2nd Edition. New York: McGraw Hill.

Todd, W.P., and G. Vittori.Texas Guide to Rainwater Harvesting, 2nd ed. Texas Water Development Board. Austin, Texas. 1997.

U.S. Department of Housing and Urban Development. Residential Water Conservation Projects. 1984. Summary Report. Report No. HUD-PDR-903. Washington, D.C.: Prepared by Brown and Caldwell Consulting Engineers for the Office of Policy Development and Research.

U.S. Environmental Protection Agency. 1998.*Water Conservation Plan Guidelines.* Office of Water. EPA-832-D-98-001. Available at http://www.epa.gov/owm/waterefficiency/

Vickers, A. 2001. *Handbook of Water Use and Conservation.* Amherst, Mass.: Water Plow Press.

Vickers, A. 2008. *Are Water Managers Becoming Lawn. Irrigation Managers?* Proceeding from Sustainable Water Sources Conference. Denver, Colo.: AWWA. http://www.awwa.org/files/Resources/Waterwiser/references/PDFs/sustainable2008_tue5-1.pdf.

Washington Department of Ecology. 1991. *Water Conservation Planning Handbook for Public Water Systems,* Publication No. 91-39. Olympia, Wash.

Whitcomb, J. 2005. *Florida Water Rates Evaluation Of Single-Family Homes for the Southwest Florida Water Management District, St. Johns River Water Management District, South Florida Water Management District, and Northwest Florida Water Management District.* Brooksville, Fla.

Index

Note: *f.* indicates figure; *t.* indicates table.

This page intentionally blank.